A l to

MAMMALS

OF AUSTRALIA

Ronald Strahan
The Australian Museum

NEW
HOLLAND

First published in the UK in 1995 by
New Holland (Publishers) Ltd
24 Nutford Place
London W1H 6DQ
London • Cape Town • Sydney

Reprinted 1997

Editor J.Greig

ISBN 1 85368 583 6

Typeset by Eps&m
Reproduction by Hirt & Carter
Printed and bound by Tien Wah Press (Pte) Ltd

Front cover: Herbert River Ringtail-Possum (H. & J. Beste)
Back cover: Feathertail Glider (G.A. Hoye)
Spine: Eastern Pygmy Possum (Dick Whitford)
Title page: Long-tailed Pygmy Possum (Dick Whitford)

Contents

Introduction

Australia is famous for its kangaroos, Koala and other marsupials. It is also home to the Platypus and Short-beaked Echidna, which are egg-laying mammals. Otherwise its representation of the diversity of the world's terrestrial mammals is remarkably low. Of the 17 major groups (orders) of mammals, only two others – the bats and the rodents – occur in its basically native fauna. There is a popular belief that marsupials are restricted to Australia but close relatives of the Australian species are found from the eastern end of the Indonesian Archipelago (east of Bali), through New Guinea to the Solomon Islands. Much more distantly related species occur from Patagonia to the Isthmus of Panama and one marsupial is common in North America. The Platypus is restricted to the eastern seaboard of Australia but the related Short-beaked Echidna extends into the lowlands of New Guinea and the rare and little-known Long-beaked Echidna lives in the mist forests of the New Guinea Highlands.

This strange distribution is not easily understood in terms of the present geography of this planet but a plausible explanation can be offered in terms of the movements of continents over very long periods. About 160 million years ago, the land surface of the world was concentrated into two masses, a northern Laurasia and a southern Gondwanaland, comprising the elements that later separated into Africa, Madagascar, India, Australia, Antarctica, New Zealand and South America. These gradually separated – India moving into the northern hemisphere to collide with Asia – but South America, Antarctica and Australia remained connected until about 45 million years ago. Before that time, South America had a diverse fauna of marsupials and at least one platypus-like monotreme, so it seems likely that representatives of these groups moved from Patagonia, across a much warmer Antarctica, into Australia.

When the Australian landmass (including what is now New Guinea) finally separated from Antarctica, the only mammals on the continent were marsupials and monotremes. From that time, until about 20 million years ago, they had the continent to themselves and, given a wealth of opportunities, these pioneers provided the basis of an enormous evolutionary radiation.

The end results of the monotreme radiation are the semi-aquatic Platypus, feeding on invertebrates at the bottom of rivers and lakes; and the decidedly terrestrial, spiny-furred echidnas, one specialised for a diet of ants, the other feeding mainly on worms in tropical leaf-litter. There may well have been other monotremes but fossil remains are so sparse that this question must be left open.

The diversity of living marsupials provides evidence of an extensive radiation. They include large and small carnivores, insectivores (including the Numbat which feeds on termites), browsers, grazers and nectar-feeders. There are fast runners, hopping species, burrowers, climbers and gliders. Habitats exploited range from tropical rainforest to stony deserts and alpine regions but, interestingly, there is no Australian aquatic marsupial equivalent to the Yapock of South America. In size, marsupials range from some that are small enough to fit into a matchbox, others weigh more than 60kg. As recently as 50 000 years ago, there were herbivorous species the size of hippopotamus.

Some 15 million to 20 million years ago, the Australian landmass had moved so far northward that its continental shelf collided with that of Asia. It was then possible for bats to fly in from Indonesia into New Guinea and Australia. Somewhat later, a few species of rodents reached Australia, probably carried on masses of floating vegetation. All of the native rodents and bats of Australia and New Guinea are closely related to, but far less diverse than, their close relatives in South-east Asia.

The only other terrestrial mammal regarded as 'native' is the Dingo, which appears to have reached Australia some 3 000 to 4 000 years ago. Closely related to the semi-domesticated Dog of South-east Asia, it must have been introduced by humans but we do not know who these were. The Dingo did not come with the Australian Aboriginals, who arrived at least 40 000 years ago.

None of the above considerations are relevant to the distribution of the marine mammals that occur off Australian coasts. Whales and dolphins were unaffected by drifting continents. Seals and their kin, which have to come ashore to breed, take advantage of appropriate coasts but those that are found in Australia have an essentially circumpolar distribution. Only one species is currently restricted to Australia.

In contrast to other continents, Australia has very few big animals. Largest are the kangaroos and wallaroos, some of which stand up to 2m tall, but most are no larger than a cat and many are of the size of rats or mice.

Classification of Australian mammals

Australia is the only continent with representatives of the three subclasses of living mammals. The only living members of the Prototheria are the monotremes, which are characterized by laying eggs. The Marsupialia give birth to minute young which lack fully developed hindlegs and attach themselves to a teat until the body is fully formed. Teats are usually in a pouch but not all marsupials have this structure. The Eutheria comprises the familiar mammals. Thanks to an efficient placenta, female eutherians are able to produce quite large young (calves, ponies) but many give birth to small infants that are blind and naked.

SUBCLASS MONOTREMATA
The two Australian members of this group, the Platypus (p.16) and Short-beaked Echidna (p.17), are easily recognized.

SUBCLASS MARSUPIALIA
SUBORDER DASYUROMORPHIA

FAMILY DASYURIDAE
The dasyurids, probably the least specialized of the Australian marsupials, are insectivorous or carnivorous and are characterized by possession of eight needle-like incisor teeth in the upper jaw and six in the lower jaw. The pouch, when present, opens ventrally. The largest member of the group, the Tasmanian Devil (p.30) is essentially a scavenger. Next in size are the mainly arboreal quolls (pp.23-25) and phascogales (p.27), which are fierce predators. Terrestrial dasyurids include antechinuses (pp.18-20), the Mulgara (p.21), the Kaluta (p.22), the Kowari (p.22), the ningauis (p.26) the dibblers (p.27), planigales (p.28), pseudantechinuses (p.29) and dunnarts (pp.31-35). All of these are rather similar in body form.

The Kultarr (p.35) resembles a dunnart in its head and body but is unusual in having long hindlegs and a very long, slender tail. Unlike other dasyurids, it has a bounding gait.

(The recently extinct Thylacine is the only known member of the family Thylacinidae.)

SUPERFAMILY NOTORYCTEMORPHIA
The one (or two) marsupial moles (p. 44) are so highly specialized for life below desert sands that their relationships are uncertain.

SUBORDER PERAMELOMORPHIA
SUPERFAMILY PERAMELOIDEA
Members of this group, comprising the bandicoots and Bilby (pp. 37-43), have a dentition similar to that of dasyurids (eight or 10 stumpy upper incisors in the upper jaw, six or eight below). They are omnivorous, feeding on the subterranean parts of succulent plants and burrowing insects: the forefeet have strong, nail-like claws. Associated with a bounding gait, the hindlegs are much larger than the forelegs and the hindfoot has a long, powerful fourth toe. The second and third toes of the hindfoot are fused together to form a single digit with two claws, used in grooming the fur, which is bristly in bandicoots, silky in the Bilby. The pouch opens to the rear.

SUBORDER DIPROTODONTA
This group, which includes the most familiar marsupials, is characterized by the presence of no more than two functional incisors in the lower jaw: these are strongly built and have chisel-like tips. The number of upper incisors varies from two to six. The second and third toes of the hindfoot are fused together to form a single digit with two claws, employed in grooming the fur. The pouch opens forwardly (except in the Koala and Wombat).

SUPERFAMILY VOMBATOIDEA
FAMILY PHASCOLARCTIDAE
The Koala (p. 47), only member of this group, is characterized by the virtual absence of a tail; it has a short muzzle, woolly fur, and large, strongly clawed feet.

FAMILY VOMBATIDAE
The three species of wombat have broad, truncated snouts, are heavily built, with short powerful limbs and stout feet with which they excavate long burrows.

SUPERFAMILY PHALANGEROIDEA
FAMILY PETAURIDAE
This group comprises the smaller, furry-tailed gliding marsupials of the genus *Petaurus* (pp. 56-58); the closely related but non-gliding Leadbeater's Possum (p. 55); and Striped Possum (p. 55). A characteristic of the gliding species is that the gliding membrane extends from the wrist to the ankle. The tail is weakly prehensile. Among its unique features, the Striped Possum has a very long, strongly clawed fourth finger.

FAMILY PSEUDOCHEIRIDAE
Pseudocheirids include a group of closely related arboreal ringtail possums (pp. 59-61) characterized by possession of a lightly furred

and very prehensile tail. Basically leaf-eaters, they have multi-cusped, grinding cheek-teeth. The Green Ringtail Possum (p. 61) has unusual pigmentation that confers a green colour to the fur. The Greater Glider (p. 59), largest of the volplaning marsupials, has a very long, non-prehensile tail and a gliding membrane that extends from the elbows to the ankles.

FAMILY PHALANGERIDAE
This rather variable group of basically leaf-eating marsupials includes the brushtail possums (pp. 52-53), which have a furry, weakly prehensile tail; the Scaly-tailed Possum (p. 54) which has an almost naked, prehensile tail; and the Cuscus or phalangers (p. 51), which have a very short snout and a very prehensile tail.

FAMILY BURRAMYIDAE
All but one of the pygmy-possums (pp. 49, 50) are mouse-sized arboreal marsupials with long, slender, prehensile tails. They feed on nectar, pollen and insects. The basically terrestrial Mountain Pygmy-possum (p. 48) is essentially terrestrial and adapted to living below the snow during winter.

FAMILY ACROBATIDAE
The single Australian species in this family (p. 62) is the smallest gliding marsupial, readily identified by its feather-like tail.

FAMILY TARSIPEDIDAE
The Honey-possum (p. 63), sole member of this family looks like a pygmy-possum but has a long, brush-tipped tongue with which it laps nectar. The claws on its feet are nail-like, except on the conjoined second and third digits of the hindfoot.

SUPERFAMILY MACROPODOIDEA
FAMILY POTOROIDAE
Members of this group of rat-kangaroos retain some possum-like features, including a moderately prehensile tail, used to carry nesting material. The Musky Rat-Kangaroo (p. 64) is unique among macropodoids in retaining the first digit of the hindfoot. Bettongs (pp. 65-66) have a short, broad head. Potoroos (p. 67) have a longer, more tapered snout. All rat-kangaroos build nests.

FAMILY MACROPODIDAE
Most numerous of the Australian marsupial families, this includes the tree-kangaroos (p. 68), hare-wallabies (p. 69), wallaroos (p. 69), nailtail wallabies (p. 76) rock-wallabies (pp. 77-79) Quokka (p. 80), pademelons (pp. 80-81) and Swamp Wallaby (p. 82). All, except the tree-kangaroos, are minor variations of a basic kangaroo pattern, with very large hindfeet extending into a large, powerful fourth toe. The tail is never prehensile. The feet are broader and shorter in rock-wallabies. Tree-kangaroos, which are secondarily arboreal, differ from typical kangaroos in having longer, more powerful forelimbs and shorter hindlimbs and hindfeet.

ORDER RODENTIA
Rodents are by far the most numerous of all mammals in terms of some 2 000 species and in numbers of individuals. They are characterized by possession of two large incisors in each jaw. These teeth have chisel-edges that work against each other in such a way that they sharpen each other as they grow continuously from the roots. The ability to gnaw into tough plant material underlies their

success. Of the 24 families of rodents, only one, the Muridae, is represented in Australia, indicative of the accidental migration of these animals into Australia.

FAMILY MURIDAE

Classification of the rodents is based on so many arcane characters that no amateur can be given guidance on this matter. It is sufficient to say that any rodent found in Australia is a murid. Nevertheless, two sub-groups can be distinguished. The 'old endemics', descended from the first rodents to reach Australia, are members of the subfamily Hydromyinae and include the Tree-rats (p. 84), Water-Rat (p. 84), False Water-Rat (p. 85), short-tailed mice (p. 85), Greater Stick-nest Rat (p.86), Broad-toothed Rat (p. 86), melomyses (pp. 87-88), hopping-mice (pp. 91-92), Prehensile-tailed Rat (p. 93), Australian mice of the genus *Pseudomys* (pp. 93-97), and rock-rats (p. 104).

What are known as the 'new endemics', descended from ancestors which appear to have reached Australia no earlier than a million years ago, are members of the world-wide subfamily Murinae. They include the native species of *Rattus* (pp. 98-102). Two species of *Rattus* are introduced, as is the House Mouse (p. 90). Because of its success in integrating into the native fauna, the House Mouse is included in this book.

It is hard to identify native rodents below the generic level; professional mammalogists often have great difficulty in determining species.

ORDER CHIROPTERA

Bats fly by means of wings that consist of thin membranes of skin that extend between the fingers, to the knees, and often between the hindlimbs and the tail. The portion between the hindlimbs is known as the interfemoral membrane. Because of the involvement of the hand in the wing, a bat has greater difficulty than most mammals in moving about on four limbs, but many can scurry quite rapidly on the ground or over the wall of a cave. When resting, most bats hang upside-down, suspended by their hindfeet, usually in the shelter of a cave, rock cleft, tree-hole, or similar humid space. Bats adapted to cold climates often hibernate in winter. The order Chiroptera is sharply divided into two suborders: the herbivorous Megachiroptera or megabats; and the insectivorous Microchiroptera or microbats. Compared with other parts of the world, Australia does not have a very diverse bat fauna; all are closely related to those of South-east Asia.

SUBORDER MEGACHIROPTERA

Megabats have relatively large eyes, uncomplicated ears and a rather long, tapered snout. The tail is short or absent and there is no interfemoral membrane. The Australian species comprise the fruit-eating flying-foxes (pp. 108-110), Bare-backed Fruit-Bat (p. 105), the nectar-feeding blossom-bats (pp. 106-107) and Queensland Tube-nosed Bat (p. 106). Flying-foxes, which are the largest of the megabats, roost in trees, often in vast numbers, and are the only bats likely to be seen in the open by day.

SUBORDER MICROCHIROPTERA

Almost one-fifth of the species of living mammals are microbats. Australian species differ from megabats in generally being much smaller (although the Ghost Bat, with a head and body length up to 13cm, is larger than the nectar-feeding megabats). With the exception of the Ghost Bat, microbats have small eyes, usually a short snout (sometimes remarkably so), and complicated ears. Some, such as the horseshoe-bats, have a complicated structure, the nose-leaf, on the snout. Most microbats use pulses of ultrasound in echolocation to form an image of their surroundings, enabling them to navigate through complex environments and to locate flying insect prey. Six families are represented in Australia.

FAMILY EMBALLONURIDAE

The sheathtail-bats are characterized by a tail that appears to pierce the interfemoral membrane from below, protruding above it. This arrangement permits greater freedom of movement of the hindlegs, enabling a sheathtail-bat to scurry with agility over the ground or the wall of a cave. The snout is tapered and there is no nose-leaf. The wings are long and narrow and the tips are folded back over the rest of the wing when an individual is at rest.

The Australian genera are *Saccolaimus* (p. 117), which lack a wing-pouch and have a deep groove in the lower lip; and *Taphozous* (pp. 115-116), which has a wing-pouch and, usually a glandular pouch on the throat, particularly in the male.

FAMILY MEGADERMATIDAE

The Ghost Bat (p. 111), only Australian member of this family, is characterized by its relatively large size, large ears, joined together at their bases, and a simple nose-leaf. It is the only Australian bat to prey on other mammals, including small bats, marsupials and rodents.

FAMILY MOLOSSIDAE

Members of this group (pp. 118-119) are known as mastiff-bats in reference to the short, wrinkled muzzle, or as free-tail bats in reference to the extent of the tail that protrudes beyond the rather small interfemoral membrane. As in the sheathtail-bats, this arrangement permits a considerable degree of freedom to the well-developed hindlimbs, leading to another common name for this group, 'scurrying-bats'. The snout is short but tapered and slightly up-tilted. The ears are large but simple. The eyes are relatively large for microbats.

FAMILY RHINOLOPHIDAE

All 70 species in this widespread genus are placed in the genus *Rhinolophus*, comprising the 'true' horseshoe-bats (distinct from the family Hipposideridae, members of which were long known as the 'horseshoe-bat'). Species of *Rhinolophus* (pp. 112-113) have large ears and very compressed snout, bearing a complex nose-leaf with a central vertical component, the 'lancet'. The toes have three joints.

FAMILY HIPPOSIDERIDAE

The leafnosed-bats in this family are very similar to horseshoe-bats (and were once known as such). They differ from horseshoe-bats in lacking a central 'lancet' in the nose-leaf and in having only two joints in the toes. The genus *Rhinonicteris* (p. 115), which has only one species, is similar in most respects to other leaf-nosed bats but is unique in that all members are orange-coloured and in being restricted to Australia. Species of *Hipposideros* (p. 114) include orange-coloured individuals.

FAMILY VESPERTILIONIDAE

This enormous family, which has representatives on all continents, includes a large number of species that differ from each other in rather minor characteristics. The Golden-tipped Bat has a notably domed cranium and coarse fur, tipped with gold. Its upper canine teeth are long, grooved and dagger-like. Wattled-bats of the genus *Chalinolobus* (pp. 120-121) and pied bats (p. 121) are characterized by lobes (wattles) on either side of the lower lip. Bent-wing bats of the genus *Miniopterus* (pp. 122-123), have long wings and a third finger which is flexed forwards when the bat is at rest. The Tube-nosed Insectivorous Bat (p. 123), which is rare in Australia, is recognizable by its tubular nostrils and its habit of resting with its wings held out from the body like a partly extended umbrella. *Myotis* is a widespread genus of fishing bats, with one species, the Large-footed Myotis (p. 124) in Australia: it flies over the surface of fresh water, catching small fishes and aquatic insects with its rake-like hindfeet. The broad-nosed bats of the genus *Scoteanax* (p. 124) have been variously classified as pipistrelles and eptesicuses. Difficult to define except in the absence of distinguishing features, they are small, rapidly flying bats with a short snout, minute eyes and no nose-leaf. Identification often depends on details of the dentition and skull.

ORDER CARNIVORA

This group, which includes lions, tigers, bears and wolves, has so few representatives in Australia that it does not need to be defined here. Two sub-groups are represented.

FAMILY CANIDAE

One member of this group is 'native' (after some 3 000 – 4 000 years' residence) in Australia. The Dingo (p. 127) is inseparable from the domestic Dog on external characteristics. Dogs and Dingoes interbreed freely.

FAMILY OTARIIDAE

Fur-seals and Sea-lions are characterized by the possession of small ears and hindlegs that can move independently and contribute to locomotion on land. ('True' seals lack external ears and have their hindlimbs directed backwards.) Two species of fur-seals (p. 128) come ashore to breed on Australian beaches. They have a thick under-fur and a somewhat shaggy appearance. The Australian Sea-lion (p. 129) has a much sleeker appearance.

ORDER SIRENIA

Although it is entirely aquatic and essentially marine, the Dugong (p. 130) is included here because it frequents shallow coastal waters and estuaries to feed on sea-grasses. It is described in the text.

Environments

Compared with the other continents, Australia is flat and dry. The only significant mountains comprise the Great Dividing Range, close to the eastern seaboard, defining a well-watered but narrow coastal plain. Westward of the range, the land becomes increasingly arid and merges into desert that extends to much of the western coast. However, the extreme south-western corner of the continent and much of Tasmania have sufficient rainfall to support tall forests and the extreme northern parts, under monsoonal influence, are seasonally wet. The nature and distribution of plants –

from trees to grasses – is largely determined by rainfall but is also limited by soils, which are shallow and infertile over most of the continent; and by temperature, which ranges from tropical to extremely cool in southern Tasmania and the tiny area of the Australian 'alps'. Botanists recognize at least 34 distinct vegetational associations in Australia, ranging from small areas of tropical rainforest in northeastern Queensland to sparse hummock grassland and arid woodland which together occupy about half of the continent. To a large extent, the distribution of Australian mammals is related to the environments that are defined by these differences in vegetation. Herbivorous species are adapted to feeding on particular plants. Insectivorous and carnivorous species feed on smaller herbivores that are themselves dependent upon the local vegetation. Survival of many arboreal species requires access to treeholes for nests. In addition to these limitations, some bats require access to caves for shelter and some terrestrial rodents and marsupials live only in rocky areas.

Taking an overall view of the distribution of mammals, it is sufficient to consider only seven major terrestrial environments.

Tropical rainforest

Rainforest can be defined as an environment dominated by tall, broad-leafed trees, forming a continuous canopy that cuts off so much light that there is little or no undergrowth: it depends upon heavy annual rainfall. Australia has some areas of temperate rainforest but the mammal fauna of these environments does not differ very much from that of adjacent wet sclerophyll forest (see below). In recent geological times, Australia has become drier and tropical rainforest is now reduced to some small areas in northern Queensland between the Great Dividing Range and the sea. Most of the indigenous mammals are arboreal and herbivorous, including bats and rodents, ringtail possums, cuscuses and tree-kangaroos. There are few ground-dwelling species and, among these, the Musky Rat-Kangaroo is of great interest. Reduced considerably in historic times by logging, this environment is now at a critical stage. Further reduction in area may put many species at great risk.

Wet sclerophyll forest

'Sclerophyll,' meaning 'hard-leaf', refers to a forest dominated by tall eucalypts that may develop a complete or almost complete canopy, but this is seldom so opaque as to preclude the development of undergrowth. This environment provides habitats for a number of herbivorous and/or insectivorous possums, arboreal and terrestrial insectivorous and carnivorous marsupials and a number of rodents and bats. Wet sclerophyll forest is mainly confined to the eastern coastal plain and the extreme south-western part of the continent and has been reduced by logging and clearing for agriculture. In consequence, many species in this environment have suffered a reduction in range, but few have become extinct.

Dry sclerophyll forest

The limits of this environment are not easy to define. In general, it begins on the western edge of the Great Dividing Range and the eastern edge of the wet sclerophyll forest of southwestern Australia, grading into woodland. It is characterized by the dominance of eucalypts that are usually less then 30m tall and do not create a complete canopy, thus permitting the growth of a scrubby understorey. Much of the original dry sclerophyll areas of Australia have been cleared for agriculture or grazing, to the detriment of many mammals that depend upon this environment.

11

Woodland
This category includes a wide range of vegetation types, in semi-arid to arid areas, characterized by scattered, low trees over shrubs or grasses. Most of it has been cleared for agriculture or altered by the introduction of grazing animals and it is here that the greatest rate of extinction of Australian mammals has occurred. The causes of these extinctions have not yet been ranked in importance but include loss of habitat; competition for food with introduced herbivores such as sheep, cattle, Goat, the Rabbit and House Mouse; and predation by Fox and Cat.

Grasslands
Much of what is termed 'grassland' in Australia would elsewhere be called 'desert'. The central, arid half of Australia has a low, scattered cover of *Triodia*, a spiky- leaved grass that grows in separate hummocks and is commonly referred to as spinifex (although, strictly speaking, spinifex is a coastal plant). Hummock grassland provides a micro-habitat and refuge for many small mammals but is not a significant source of food.

Tussock grassland, in the somewhat less arid parts of the continent, is characterized by clumps of *Astrebla*, or Mitchell Grass, which provides a micro-habitat for many small native mammals but does not contribute much to their diet. On the other hand, it is palatable to cattle and grazing of these animals has probably been to the detriment of native species.

Heathland
Heaths comprise extremely dense, low vegetation of plants with hard or prickly leaves. They are usually coastal and adapted to strong, salt-laden winds. Heathland also occurs in high regions of the Great Dividing Range (the 'alps') and in high, inland parts of Tasmania. A few small native mammals are adapted to these habitats.

Mangroves
Mangroves create dense, low vegetation on the edges of coastal or estuarine beaches. Only one terrestrial mammal, the False Water-Rat, is dependent upon this environment but it is utilised by many flying-foxes.

How to use this book

A little more than half of the 294 living species of Australian mammals are described in this book, selected on three general criteria: first that every genus be included; second that the species selected should indicate the range of variation within each genus; thirdly that, other considerations being equal, preference be given to those species most likely to be encountered.

Given an adequate look at an animal, it should be possible, by reference to the silhouettes (p.15) to restrict it to one or a few families; in some instances, the silhouette may even serve to identify the genus or species. Points to look for include the shape of the head, size and shape of ears, body proportions, shape and length of tail and whether it is tufted.

An initial broad identification can then be tested by reference to the Classification (pp. 5-10). Indeed, this section should be read and re-read until the major categories of the mammal fauna are understood: this provides the foundation on which the species accounts are constructed.

Having narrowed down the possibilities, the reader should then consult the relevant photographs and text to confirm the identity of the genus. The characteristics of each genus are mentioned briefly in the opening sentences of the first of its species to be mentioned. Thus, in dealing with the planigales, the first species to be dealt with in the text is the Common Planigale, *Planigale maculata*, (p. 28). The genus *Planigale* is briefly defined at the beginning of this species' account. Listed below the description are those species of the genus that are **not** treated.

Identification of many of the smaller marsupials, rodents and bats is difficult, even with an animal in the hand. Confirmation often requires reference to details of the skull or dentition – in some cases to elaborate chemical techniques, all of which are matters for a professional mammalogist. However, most of the middle-sized to large marsupials can be identified at a distance with reasonable certainty by non-specialists. The amateur who can determine the genus of a small Australian mammal has reason to be satisfied. If he or she can hit upon the species, there is cause for self-congratulation.

Only two measurements are given for terrestrial mammals: the head-and-body length (HB), measured from the tip of the snout to the base of the tail and tail-length (T), measured from the base of the tail to the tip. Only the total length (TL) is given for marine mammals. Measurements are given as a range for adults and sub-adults.

Distribution maps show the *maximum* range of the species. This does not imply that it is to be found in every part of that range, only that it is likely to occur **in appropriate habitats within the range**.

On the other hand, it is very unlikely that a species will be found outside the distribution indicated on its map: a tentative identification that indicates this is probably incorrect.

Abbreviations

HB   Head and Body length (measure from the tip of the snout to the base of the tail)

T   Tail length (measure from the base of the tail to the tip)

TL   Total Length (applied to marine animals only)

Mammal watching

Other than the Numbat, Musky Rat-Kangaroo, Sea-lions and Fur-seals, Australian mammals sleep during the day. Some kangaroos and wallabies become active in the late afternoon but, otherwise, one must look for mammals at night. Many arboreal species can be located by a spotlight. They are first located by the light reflected from their eyes and, since most remain motionless while the beam is held on them, it is often possible to approach close enough to make an identification. Apart from some kangaroos and larger wallabies, few terrestrial species 'freeze' in spotlight, so there is little change of finding them except by trapping.

State and Territory laws restrict the trapping of native mammals to holders of permits issued for research or survey purposes. The keen mammal watcher should therefore make contact with a mammalogist engaged in fieldwork, a National Park Ranger, or the mammal survey group of a zoological or field naturalists' society of sufficient status to hold a trapping permit. As an unobtrusive observer or unpaid labourer, one can then assist in setting traps and pitfalls, being rewarded by the opportunity to examine living animals closely after they have been identified and before they are released. Occasionally, under expert guidance, one may locate the burrow of a small marsupial or rodent and, remaining very quiet, observe animals emerging to forage.

Without long experience, it is very hard to identify bats in flight, but flying-foxes can be distinguished from other bats by their greater size and most can be identified to species level when seen hanging from the branches of trees in their daytime roosts. Small bats (Microchiroptera) can often be identified in flight by their ultrasonic calls but special electronic devices are required to translate these into frequencies that are audible to humans. Absolute identification requires trapping, either in coarse-meshed nets of very fine thread (mist nets) or in 'harps,' which are large frames supporting tightly strung fine, vertical piano wires. Unable to detect the wires, bats fly into these, are momentarily stunned, and fall into a bag at the bottom of the frame. Needless to say, the use of mist nets and harps is strictly controlled by fauna authorities.

Amateurs sometimes encounter bats in caves. If bats are seen to leave a cave regularly in the evening, it is unlikely that much harm would be caused by the entry of humans into that cave by day when the bats are roosting. Nevertheless, frequent intrusion may lead a colony to vacate its cave, with a consequent reduction in the population of the species. The situation is much worse with those bats that hibernate in caves in the southern-half of Australia in the colder part of the year. Each time such a colony is disturbed, the bats use up part of the energy reserves (fat) stored in the body to carry them through until activity and feeding resumes in the spring. Even infrequent disturbances (perhaps as few as two or three occasions) that arouse the bats to panic flight can cause exhaustion of food reserves before spring arrives, leading to deaths. One should not enter a cave housing an active colony except under the guidance of an expert, and never enter a cave that is home to a hibernating colony.

Key to Symbols

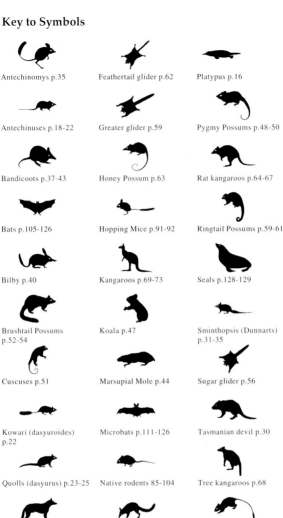

Antechinomys p.35	Feathertail glider p.62	Platypus p.16
Antechinuses p.18-22	Greater glider p.59	Pygmy Possums p.48-50
Bandicoots p.37-43	Honey Possum p.63	Rat kangaroos p.64-67
Bats p.105-126	Hopping Mice p.91-92	Ringtail Possums p.59-61
Bilby p.40	Kangaroos p.69-73	Seals p.128-129
Brushtail Possums p.52-54	Koala p.47	Sminthopsis (Dunnarts) p.31-35
Cuscuses p.51	Marsupial Mole p.44	Sugar glider p.56
Kowari (dasyuroides) p.22	Microbats p.111-126	Tasmanian devil p.30
Quolls (dasyurus) p.23-25	Native rodents 85-104	Tree kangaroos p.68
Dingo p.127	Numbat p.36	Tree rat p.84
Dugong p.130	Phascogales p.27	Wallabies p.69, 71-83
Echidna p.17	Planigales p.28	Wombats p.45-46

15

Platypus *Ornithorhynchus anatinus* HB 300-420mm T 100-130mm

H. & J. Beste

A. Young

Platypus hindfoot, showing spur

The Platypus is seldom seen except when resting at the water surface. It can be identified by its duck-like beak, domed head, white fur around the eyes, outspread limbs and broad, flattened tail. One of the three living species of egg-laying mammals (order Monotremata), it is amphibious, swimming with its webbed forefeet and using the webbed hindfeet to steer or to brake. Its nostrils are towards the tip of a leathery bill, shaped like that of a duck. When submerged to feed on invertebrates at the bottom of a river or pool, the eyes and nostrils are closed and food is located by a combination of touch and an electric sense. During the day a Platypus rests in a long burrow in a bank, emerging in the evening (sometimes by day under cold conditions) to feed and to interact with other individuals. The male is territorial, defending a stretch of river bank and adjacent water. The hollow spur on the ankle of the male, connected to a venom gland, appears to be employed in aggression between males. Mating begins in August in the warmer part of the range, as late as October in the colder part. Two leathery-shelled eggs are incubated by the female for about two weeks in a specially constructed breeding burrow. The young are suckled for four to five months.

Short-beaked Echidna *Tachyglossus aculeatus* HB 330-500mm T 37-50mm

G.B. Baker

This species is recognized by a covering of spines and a tubular snout which accommodates a long, mobile tongue used to lap up ants. Related to the much larger Long-beaked Echidna of New Guinea, it is the only echidna found in Australia, occurring over most of the continent and in Tasmania. It feeds around evening and dawn (during the day under cold conditions), mostly upon ants, which it exposes from their nest with powerfully clawed forelimbs. When disturbed, it may lodge itself in a crevice by powerful exertion of its limbs and spines, curl into a ball, or lower itself, horizontally, into the soil by excavating movements of all four limbs. Mating occurs in July and August and the single egg is incubated for about 10 days in a pouch on the mother's abdomen. The young which lacks fully developed hindlimbs at the time of hatching, is suckled for at least three months, first in the mother's pouch and later in a burrow, where it is left while the mother is foraging. The juvenile follows the mother after weaning and does not attain adulthood until the age of about 12 months. The species shows considerable geographic variation: in warmer climates, the hair is barely visible between the spines; in colder regions, the hair may be almost as long as the spines.

Yellow-footed Antechinus *Antechinus flavipes* HB 90-170mm T 60-150mm

Babs & Bert Wells

The genus *Antechinus* with some 10 species in Australia and New Guinea, is characterized by a sharply tapered snout, rather short ears and a fully (though sparsely) furred tail. The hindfoot has five digits and transversely striated pads. One of the most successful species, the Yellow-footed Antechinus, varies in colour but is distinguished by a combination of pale eye-rings, a black-tipped tail, and a distinction between the slate-grey colour of the head, the orange-brown under parts (paler in the southern and western parts of the range). It lives in coastal forests and woodland from tropical Queensland to southern Victoria where it is an active nocturnal predator on small rodents, birds and insects, and also feeds on flowers and nectar. Although mainly terrestrial, it chases prey into low vegetation. Mating occurs from late winter to spring, later in the southern part of the range. As many as eight young are carried on the female's eight teats in a rudimentary pouch for up to five weeks. Thereafter, they are suckled in a nest until about three months old, sharing the nest with their mother until about 10 months old. The male mates when about 12 months old and dies shortly after this. Three subspecies are recognized, one in tropical Queensland, one in south-eastern Australia and one in the south-west.

Species of *Antechinus* not dealt with: *A. bellus, A. godmani, A. leo.*

18

Swamp Antechinus *Antechinus minimus* HB 90-140mm T 60-100mm

J.H. Seebeck

The Swamp Antechinus differs from the other three antechinuses which occur within its range by its long foreclaws, small eyes, short ears, grey head and shoulders, rich yellowish brown rump and paler underparts. It feeds on small arthropods in cool wet heathland and tussock grassland. Mating occurs from May to August. The pouch is rudimentary with six to eight teats, all usually occupied after mating. Detaching after two months, the young are suckled in a nest until about three months old. The male mates when 12 months old and dies soon after.

Brown Antechinus *Antechinus stuartii* HB 70-140mm T 60-110mm

G. Little

This common marsupial carnivore is characterized by dullish colouration, greyish-brown above, paler below, lack of eye-rings and a moderately hairy, thin tail, about as long as the head and body. Its habitat is the leaf-litter of eucalypt forests. Although usually terrestrial, it ascends shrubs and low trees in pursuit of arthropod prey. In northern Queensland it may nest arboreally. It is a nocturnal predator on insects and other arthropods. It sleeps during the day, sometimes in communal nests. Mating occurs from August in the south to September in the north. The male dies shortly after mating.

Dusky Antechinus *Antechinus swainsonii* HB 90-190mm T 70-120mm

G. Hoye

Probably the most frequently trapped or observed of all the antechinuses, this species, which shares its distribution with the Brown and Swamp Antechinus, is distinguished from other antechinuses within its range by small eyes and ears, long foreclaws, and soft, deep brown or grizzled grey-brown fur on the upper parts, pale grey below. It is a nocturnal, terrestrial predator on insects and other arthropods in the leaf-litter of wet forests and heathlands, also eating some plant material such as blackberries. During the day it rests in a spherical nest in a short burrow excavated into a creek bank. Mating occurs in midsummer over most of the range (later in colder climates). The female has a rudimentary pouch and usually carries young on all six to eight teats until about eight weeks old. Thereafter, they cling to the mother's fur for five to six weeks while being suckled in the nest, becoming more or less independent when about 12 weeks old. The male dies shortly after mating. It is one of the first of the dasyurids to colonize an area recovering from extensive bushfire, feeding largely upon the House Mouse, which is usually the first herbivore in such areas.

Mulgara *Dasycercus cristicauda* HB 120-220mm T 70-130mm

G.B. Baker

Restricted to the sandy inland deserts, this medium-sized dasyurid, the only member of its genus, is sandy brown above, lighter below, and distinguished by its size, robust structure, larger eyes and crest (not brush) of black hairs on the distal two-thirds of the tail: the ears are short and rounded. It is a nocturnal predator upon large invertebrates and small vertebrates. During the day it rests in a burrow system, often with a number of pop-holes, but may emerge to bask in the sun. It is so adapted to desert existence that it does not need to drink, obtaining all necessary water from its prey. Mating is influenced by rainfall but is usually from March to April, although females with young have been taken in the wild between June and December. Captive animals have mated from mid-May to mid-June. In common with some other dasyurids, the Mulgara stores excess food as fat in the base of its tail which, in well-nourished individuals, is carrot-shaped. The female usually carries four to seven young on her four to eight teats for about seven weeks in a vestigial pouch. Thereafter, they are suckled in a nest until about 14 weeks old. Populations appear to undergo considerable fluctuations.

Kaluta *Dasykaluta rosamondae* HB 90-110mm T 60-70mm

P. Wolley & D. Walsh

Once known as the Little Red Antechinus, the Kaluta differs from *Antechinus* in lacking an upper third premolar. The tail is flicked vertically when feeding and, under favourable conditions, is swollen with fat at the base. The Kaluta sleeps by day in the shelter of a tussock, emerging at night to prey upon insects and small lizards. Mating occurs in September or October. The female has eight teats and may carry up to eight young which become independent at the age of 14 to 16 weeks. Only one litter is reared in a year. The male dies shortly after mating.

Kowari *Dasyuroides byrnei* HB 130-180mm T 110-140mm

H. & J. Beste

Characterized by a brush of dark brown hair all around the terminal half of the tail, this stocky dasyurid occurs only in central desert areas. It is a nocturnal predator upon arthropods and small mammals and obtains sufficient water from its food. During the day it sleeps in a nest at the end of simple burrow. Breeding occurs through most of the year, and the female carries up to six young in a vestigial pouch for about eight weeks, becoming independent at the age of about 16 weeks.

Spotted-tailed Quoll *Dasyurus maculatus* HB 350-760mm T 340-550mm

G.B. Baker

Quolls are medium-sized carnivorous marsupials with a tapered muzzle; rather short, rounded ears and a non-prehensile tail that is well-furred and shorter than the head and body. The body fur is spotted in all species. The Spotted-tailed Quoll (or 'Tiger Cat') is the largest species. Its brown upper body fur has white spots which extend onto the tail. The hindfoot has five toes and it is equally agile on the ground or in a tree. It occupies a wide variety of habitats from rainforest to heathland. A nocturnal predator feeding on animals ranging in size from small wallabies to large insects, it is not averse to carrion and may enter camps in search of food or raid poultry pens. Individuals may become tolerant of humans. It usually sleeps by day in a den in a hollow log, cave or rock crevice. The females has a relatively well-developed pouch with six teats. Mating is usually from April to July and about five young are carried on the teats until seven weeks old. Thereafter, they are suckled in a den. When about 10 weeks old, young may accompany their mother on foraging excursions, clinging to her fur with teeth and claws. They become independent at about 18 weeks. Its range and numbers have declined considerably since European settlement but, although it has a very disjunct distribution, it seems not to be endangered.

Western Quoll *Dasyurus geoffroii* HB 270-360mm T 200-280mm

Babs & Bert Wells

The upper parts of this rather large dasyurid are brown with small white spots, except on the tail. It is now restricted to open forest and woodland in south-western Western Australia. It has five toes on the hindfoot, associated with its ability to pursue prey in low foliage. It is a nocturnal predator, taking large insects and vertebrates, and also eats carrion. During the day it sleeps in a shallow burrow or equivalent space. Mating is mostly from May to June. The female has six teats in a well-formed pouch and gives birth to six young which become independent at about 22 to 24 weeks old.

Northern Quoll *Dasyurus hallucatus* HB 120-310mm T 200-300mm

H. & J. Beste

The smallest quoll, this species has white spots on the brown-grey upper parts. The hindfoot has five toes. It inhabits tropical to subtropical open forest, savanna and scrubland, favouring rocky country. A nocturnal predator, its diet consists of large insects, small vertebrates, and fruit. It sleeps by day in a hollow log or crevice. Mating occurs in June or July and six young are carried on the six to eight teats for about 10 weeks. They are independent after 20 weeks old.

Eastern Quoll *Dasyurus viverrinus* HB 280-450mm T 170-280mm

D. Watts

This medium-sized quoll is spotted on the body like the Western and Northern Quolls but is dimorphic, having either brown or black upper parts. It is the only quoll in Tasmania and appears to be either very rare or extinct in its original range on the south-eastern coastal mainland. It is distinguishable from the Spotted-tailed Quoll by its smaller size, absence of spots on the tail and the presence of only four toes on the hindfoot. It is a nocturnal predator upon small terrestrial mammals, birds, reptiles and insects and other large arthropods, also eating carrion, fruits and even grasses. It occupies a variety of habitats from dry sclerophyll forest to heathland and pasture. During the day it sleeps in a grass-lined den in a natural crevice, hollow log or short burrow. In Tasmania mating occurs from May to June. There are five to eight teats in a quite rudimentary pouch. The female usually carries young on each teat for 8 to 10 weeks, after which they are suckled in a den. They become independent when about 20 weeks old and, before this, may accompany the mother on her foraging excursions, clinging to her fur with teeth and claws.

Wongai Ningaui *Ningaui ridei* HB 60-70mm T 60-70mm

Dick Whitford

Ningauis resemble planigales, but have a less flattened skull and narrower hindfeet. They also resemble dunnarts but are generally smaller and have broader hindfeet. The fur is rather bristly. The Wongai Ningaui, which inhabits the central deserts, is able to climb in low vegetation, aided by its long, semi-prehensile tail. It is a nocturnal predator on arthropods. By day it sleeps under shelter. It is seldom trapped but often collected in pitfalls. Mating is probably from September to December. The female has six to seven teats in a rudimentary pouch and frequently carries five to seven young for about four weeks, becoming independent at about 13 weeks old. Two litters may be reared in a year.

Species of *Ningaui* not dealt with: *N. timealeyi, N. yvonneae.*

Southern Dibbler *Parantechinus apicalis* HB 120-140mm T 90-110mm

Babs & Bert Wells

Dibblers resemble antechinuses but differ in having second upper incisors that are equal to, or shorter than, the fourth, and third premolars that are shorter than the second. The Southern Dibbler (once known simply as the 'Dibbler') is significantly larger than the Northern Dibbler. It inhabits banksia heathland and is a nocturnal predator on small arthropods. During the day it sleeps in a nest of twigs and grass. Mating occurs around March. The female has eight teats in a rudimentary pouch and usually carries eight young, becoming independent at 16 weeks old.

Brush-tailed Phascogale *Phascogale tapoatafa* HB 160-230mm T 170-220mm

The two species of *Phascogale* are the most arboreal of the dasyurids. The lower third premolar is notably smaller than the second and the distal half of the tail bears a conspicuous black brush like that of the Red-tailed Phascogale: the proximal half is grey. The Brush-tailed Phascogale sleeps by day in a leaf-lined nest in a tree-hole and is a nocturnal arboreal and terrestrial predator upon small vertebrates and large arthropods. It inhabits dry sclerophyll forest. Mating occurs around June. The female has eight teats and three to six young are carried for about six weeks, then suckled in a nest until about 14 weeks old.

Babs & Bert Wells

Red-tailed Phascogale *Phascogale calura* HB 90-130mm T 120-150mm

Babs & Bert Wells

This arboreal dasyurid has a notable black brush on the distal half of the tail but the fur on the proximal half is reddish. It sleeps by day in a tree-hole or hollow log. At night it preys arboreally and terrestrially on insects and small vertebrates. Obtaining enough water from its food, it does not need to drink. Mating occurs in July and, in common with antechinuses, the male dies shortly after mating. The female has eight teats but lacks a pouch and carries eight young which are left in a nest after they become detached from the teats. They are suckled until becoming independent when about 18 weeks old.

27

Common Planigale *Planigale maculata* HB 70-100mm T 60-100mm

The four species of this genus are characterized by a notably flattened head, short ears and outwardly directed hindfeet when running. The Common Planigale, largest of the species, has cinnamon to grey-brown upper parts, flecked with white, and is pale below, with a white throat and chin: the tail is shorter than the head and body. A nocturnal predator on arthropods and small vertebrates, it occupies a wide variety of habitats. During the day it sleeps in a nest under fallen timber or under a rock. The females has eight to 12 teats in a well-developed pouch and often carries eight young.

Species of *Planigale* not dealt with: *P. gilesi, P. tenuirostris.*

Long-tailed Planigale *Planigale ingrami* HB 50-60mm T 50-60mm

Weighing about 4g, this is the smallest marsupial. Its slender tail is about the same length as the body and (very flattened) head. It inhabits seasonally flooded grassland and savanna on soils that crack in the dry season, sleeping during the day under a tussock, a stone, or in a soilcrack, and preying at night on insects and small reptiles. Mating is mostly from November to February and the female carries four to eight young for about six weeks in a well- developed pouch. Thereafter, they are suckled in a nest, becoming independent at about 12 weeks.

Fat-tailed Pseudantechinus *Pseudantechinus macdonnellensis* HB 90-110mm
T 70-90mm

Babs & Bert Wells

The three species of *Pseudantechinus* resemble antechinuses in many respects, but have a relatively shorter tail (usually swollen at the base with stored fat). The hindfeet are relatively broad and there are chestnut patches behind the ears. The upper third premolar is very small and there is no third premolar in the lower jaw. The Fat-tailed Pseudantechinus, which is the best-known member of the genus, is greyish-brown above and greyish-white below. It inhabits arid country, usually on rocky hills and breakaways, and spends the day in shelter under a rock or, sometimes, in a cavity in a termite mound. It may emerge during daylight hours to bask in the sun. At night it is an agile terrestrial predator, mainly upon insects but also taking small vertebrates. The male is slightly larger than the female. Mating occurs in winter or spring. Only one litter is reared in a year. The female has six teats in a moderately developed pouch and may carry up to six young which are suckled for about 14 weeks. The male and female may survive to breed for up to three years, in contrast to species of *Antechinus* in which the male dies shortly after mating.

Species of *Pseudantechinus* not dealt with: *P. ningbing,*
P. woolleyae.

29

Tasmanian Devil *Sarcophilus harrisii* HB 500-650mm T 230-260mm

L.F. Schick

Largest of the living dasyurids, this species lived in prehistoric times over much of the eastern mainland but is now restricted to Tasmania. The size of a small dog, it is black with white patches, usually on the chest, and sometimes on the rump. It is found in habitats varying from wet sclerophyll forest to woodland and scrubland and spends the day under shelter of fallen timber, in a rock crevice or hollow log. It feeds at night on mammal carcasses, weak or immobilized mammals, birds (such as penned poultry), and beetle larvae. Individuals competing for access to a carcass utter loud screams and are markedly aggressive towards each other. Its powerful jaws enable it to consume all parts of a carcass and even crush the skull of a sheep. Adults are entirely terrestrial, but the young can climb in strong shrubs and on trees with sloping trunks, enabling them to prey on the eggs, and young, of nesting birds. Mating occurs in March or April. The female has four teats in a complete, backwardly opening pouch and usually carries two to three young for 13 to 15 weeks. Thereafter, they are suckled in a den until becoming independent at the age of 28 to 30 weeks.

Fat-tailed Dunnart *Sminthopsis crassicaudata* HB 60-90mm T 40-70mm

G. Hoye

The 14 or so members of the genus *Sminthopsis* are characterized by a very pointed muzzle, large eyes and ears and third premolar teeth that are almost as large as the second. The hindfeet are proportionally much larger and narrower than in antechinuses. The Fat-tailed Dunnart has a tail that is slightly shorter than the head and body in northern populations, much shorter in the southern populations. The inter-digital pads of the hindfoot are finely granulated. It occupies cool temperate habitats from open forest, through woodland to heathland and spends the day in a nest of dry foliage under fallen timber, a stone, hollow log, or clump of grass. At night it forages in open areas, preying upon medium-sized arthropods and small reptiles. It does not need to drink. Individuals are usually solitary and occupy large, changing home ranges. In the colder part of the year several may rest together to conserve heat. In well-nourished individuals the tail is swollen at the base and distinctly carrot-shaped: this swelling diminishes in winter. Population density varies considerably, being largely determined by the environment created by the mid-successional regeneration after extensive wildfire. The male is slightly larger than the female. Mating extends from July to December. The female has eight to 10 teats and carries four to 10 young in a reasonably well-developed pouch. When about five weeks old, these are suckled in a nest to independence at nine to 10 weeks.

Species of *Sminthopsis* not dealt with: *S. aitkeni, S. archeri, S. butleri, S. dolichura, S. douglasi, S. gilberti, S. griseoventer, S. longicaudata, S. ooldea, S. psammophila.*

Hairy-footed Dunnart *Sminthopsis hirtipes* HB 70-90mm T 70-100mm

Babs & Bert Wells

Like the Lesser Hairy-footed Dunnart, this mostly inhabits sandy regions with woodland, shrubland or tussock grassland. Both species have hairs between the granules of the soles and palms and a fringe of long hairs extending outward from the sides of the long, broad feet, obvious adaptations to a sandy substrate. During the day the Hairy-footed Dunnart sleeps under shelter, often excavated by another species (e.g., bull-ants, hopping-mice). At night it is a predator on insects and small lizards. The base of the tail becomes fattened under favourable conditions. Mating appears to be in response to rain and growth of vegetation. The female has six teats in a circular pouch.

White-footed Dunnart *Sminthopsis leucopus* HB 70-120mm T 60-100mm

G.B. Baker

Distinguished from the Common Dunnart by its southern Victorian and Tasmanian range, it has light grey upper parts (darker on the rump), paler belly and a very sparsely furred tail. It occupies the floor of habitats from cool closed forest to heathland and tussock grassland. By day it sleeps in a nest under the shelter of fallen timber, or stone, or in a hollow log. It is a nocturnal predator on large insects and small reptiles and mammals. The male is larger than the female. Mating probably occurs from July to September. The female has 10 (mainland) or eight (Tasmania) teats and frequently rears eight to 10 young.

Stripe-faced Dunnart *Sminthopsis macroura* HB 70-100mm T 80-100mm

Widespread through the central and northern mainland, this dunnart has a distribution that overlaps with several other species of *Sminthopsis*. It is distinguished from all these (except the Long-tailed Dunnart) by the length of its tail, which is about 1.25 times longer then the head and body. (The tail of the Long-tailed Dunnart is more than twice the head and body length.) As in some other dunnarts, well-nourished individuals often have a tail that is swollen at the base. There is a dark stripe from between the eyes to between the ears, but this is also present in some other dunnarts. It occurs in a wide variety of arid and semi-arid habitats, but tends to favour tussock grassland. By day it sleeps under shelter or in a soil crack and frequently becomes torpid for periods of several hours, particularly during the winter. At night it is a predator on arthropods and possibly small vertebrates. Breeding in the wild is probably dependent upon unpredictable rainfall and consequent productivity but, in captivity, mating occurs from June to February and two litters can be raised in a year. The female has eight teats in a well-developed pouch and it is usual for her to carry eight young for about 40 days. These are suckled in a nest until about 10 weeks old.

Common Dunnart *Sminthopsis murina* HB 60-100mm T 70-100mm

G. Hoye

Possibly less common than the Stripe-faced Dunnart, this species is better known because its distribution brings it close to five capital cities. It is slate-grey above, greyish-white below, and can be distinguished from other dunnarts by a combination of the lack of a head-stripe; unfused, hairless, unstriated interdigital pads on the hind-feet; and a tail that is about the same length as the body and is never swollen at the base. It occupies reasonably well-watered dry sclerophyll forest, woodland and heathland but extends into sub-arid regions and is best represented in areas that are in the process of regeneration of ground vegetation one to two years after a fire. During the day it sleeps in a cup-shaped nest under cover or in a shallow burrow, sometimes becoming torpid for several hours. At night it is a terrestrial predator on medium-sized insects and spiders. The male is slightly larger than the female and is aggressive in the mating season, which occurs from July to September. Two litters are usually reared in a year. The female has eight to ten teats in a well-developed pouch and it is not uncommon for every teat to be occupied by young. At the age of about five weeks, the young are suckled in a nest, becoming independent when nine to 10 weeks old.

Red-cheeked Dunnart *Sminthopsis virginiae* HB 80-130mm T 90-140mm

Dick Whitford

This is distinguished by a very distinct dark head-stripe from the tip of the muzzle to between the ears; reddish cheeks; rather bristly, short fur and a thin, scaly tail. Inhabiting tropical savanna in Australia and New Guinea, it sleeps by day in the shelter of dense vegetation. At night it is a predator on arthropods and small lizards. In Australia mating extends from October to May. The Australian female has eight teats in a well-developed pouch, and often carry eight young. After becoming detached from the teats at eight to 10 weeks they are suckled in a saucer-shaped nest until four to six months old.

Kultarr *Antechinomys laniger* HB 70-110mm T 100-150mm

Babs & Bert Wells

The Kultarr resembles dunnarts in most aspects of its anatomy except for very long hindlegs and feet and a very long, brush-tipped tail. When moving, it bounds with its hindfeet and lands on its relatively long forefeet (differing in this respect from hopping-mice which also live in its desert habitat). It is a nocturnal predator on small arthropods, spending the day under a rock, log, clump of vegetation, or in a soil-crack. Mating is mostly between July and November and about four young are carried on the female's five to eight teats for about 30 days, then suckled in a nest until about 90 days old.

Numbat *Myrmecobius fasciatus* HB 200–280mm T 160–210mm

H. & J. Beste

Once distributed from central New South Wales to Western Australia, this species is readily distinguished by a long, pointed snout, a black eye-stripe from the snout to the ear, a long, furry tail and 10 or more transverse white stripes across the reddish-brown to black body fur. The only member of its family, it is now restricted to a small area in the south-western corner of the mainland in remnant open jarrah and wandoo forests. It is completely diurnal, sleeping at night in a nest of shredded bark under fallen timber or in a hollow log. By day it forages on the ground, using the forepaws to excavate the subsoil galleries of termites (seldom their nests), extracting these insects by very rapid and seeking movements of its long, mobile, sticky tongue. The 13 or so teeth on each side of the upper and lower jaws are simple peg-like structures. The female has four teats but no pouch. Mating occurs in January and it is usual for four young to be carried for about 20 weeks. Thereafter, they are suckled in a grass-lined burrow 1 to 2m long but may later be moved to a hollow log. Juveniles may accompany their mother for several months, sometimes clinging to her, and do not become independent until 11 to 12 months old. The decline of this species appears to be largely due to predation by the Fox.

Northern Brown Bandicoot *Isoodon macrourus* HB 300-470mm T 80-220mm

G.A. Hoye

The genus *Isoodon*, with three species, is distinguished from the other three genera of living Australian bandicoots by a shorter, less acute snout and shorter, more rounded ears. The Northern Brown Bandicoot is similar to the Southern Brown Bandicoot but has slightly browner upper fur and is larger. It occupies a wide range of habitats from subtropical and wet sclerophyll forest to woodland with dense ground cover. It has a mainly northern distribution and also occurs in New Guinea. By day it sleeps in a large nest of sticks, vegetation and earth, which appears formless but has a defined entry and exit. At night it employs its strongly clawed forefeet to dig in forest litter and soil for burrowing arthropods and succulent parts of plants; it also eats berries. The male is larger than the female and both sexes are solitary and aggressive. Mating can occur at any time of the year (except autumn in the southern part of the range). Gestation is completed in the remarkably short period of 12.5 days. The female has eight teats in a backwardly directed pouch but it is unusual for more that four of these to be occupied at a time. Young remain on the teats for about seven weeks and are suckled in the nest until about eight weeks old (when it is not unusual for the next litter to be born). The female becomes sexually mature when 12 to 16 weeks old.

Golden Bandicoot *Isoodon auratus* HB 190-300mm T 80-120mm

K. Johnson

Smallest species of its genus, this bandicoot is distinguished by shiny red-brown cheeks and flanks, flecked with black: the underparts are white. (The population on Barrow Island is much darker than the mainland population, not golden.) Formerly distributed over much of central and northern Australia, it is now restricted to a part of the Kimberley coast and Barrow Island. On the mainland it occupies habitats ranging from sand dunes with tussock grass cover, through eucalypt woodland, to tropical vine thickets. On Barrow Island it shelters during the day in a cave or under a spinifex hummock. At night it digs into the soil in search of insects such as termites, ants, caterpillars, centipedes and succulent plant material. It is also known to eat turtle eggs and small reptiles. Individuals are solitary, territorial and aggressive. Mating appears to be related to rainfall, with peaks in the wet season (December, January) and the dry season (August). On Barrow Island mating occurs throughout the year. The female has eight teats in a well-developed and backwardly opening pouch but it is unusual for more than two young to be accommodated. The cause or causes of its decline are obscure but it is known to be among the prey of the feral Cat.

Southern Brown Bandicoot *Isoodon obesulus* HB 280-360mm T 90-140mm

H & J. Beste

Similar in appearance to the Northern Brown Bandicoot, this species is smaller and tends to be black-brown above and whitish below. Except for a subspecies on Cape York Peninsula, the Southern Brown Bandicoot has a southern distribution, including Tasmania. It occupies habitats ranging from wet and dry sclerophyll forest to woodland with a dense understorey and is most prevalent when this is regenerating after fire. It sometimes occurs in suburban gardens where its presence is revealed by conical excavations in lawns. Individuals are solitary and aggressive, particularly in the mating season. The day is spent sleeping in a nest of sticks, vegetation and soil; under wet conditions this may be raised above ground level on a mound of earth. One individual may use several nests. At night it uses its strongly clawed forefeet to dig in leaf-litter and soil for burrowing arthropods and succulent parts of plants. Mating occurs from May to September and three litters may be reared in a year. The female has eight teats in a backwardly opening pouch but it is unusual for more than four of these to be occupied by young. Young detach from the teats when about seven weeks old and are suckled in the nest until independent at about nine weeks. The female become sexually mature when about three to four months old.

Bilby *Macrotis lagotis* HB 300-550mm T 200-300mm

L. Lochman

This endangered species is the only surviving member of its genus, characterized by long, rabbit-like ears, silky fur, a long snout and a tail that is white-tipped but black over much of the proximal-half. It occupies arid to semi-arid woodland, shrubland and hummock grassland, being most prevalent when this is regenerating after fire. During the day it sleeps in a nest at the bottom of a complex burrow that may be 2m deep. The entrance to the burrow is usually at the base of a tussock or shrub, or against the base of a termite mound. A burrow may be shared by a male and a female or by a female and her immature young. At night it forages for insects, fungi, succulent bulbs and tubers by digging with its strongly clawed forefeet. It also eats some fruits and seeds. It does not need to drink, obtaining enough water from its food. Although the male is solitary and territorial, it is not aggressive towards others. Mating may occur at any time of the year but it is probably influenced by rainfall. The female has eight teats in a backwardly opening pouch but it is unusual for more than two to be occupied. Young detach from the teat when 11 to 12 weeks old and are weaned when 13 to 15 weeks old. The female becomes sexually mature when about five months old. In captivity, mating is continuous.

Long-nosed Bandicoot *Perameles nasuta* HB 310-430mm T 120-160mm

Dick Whitford

 Members of the genus *Perameles* have a longer and more acutely pointed snout than in *Isoodon*. The rear half of the sole of the hindfoot is covered with hairs. The Long-nosed Bandicoot, most widely distributed of the four living species of the genus, is greyish-brown above, cream-coloured below, and distinguished from the others by the absence of distinct transverse light and dark barring over the rump. Faint barring may be present. Its habitats range from rainforest and sclerophyll forest to well-watered woodland. During the day it sleeps in an inconspicuous nest of vegetation over a shallow scrape in the soil. At night it feeds – like other bandicoots – by digging for burrowing arthropods and succulent parts of plants with its strongly clawed forepaws. It is not uncommon in suburban gardens where its presence is revealed by conical excavations made in lawns. The male and female are both territorial and aggressive. Breeding occurs throughout the year but less in winter. Gestation takes only 12.5 days. The female has eight teats in a well-developed backwardly opening pouch but seldom carries more than two or three young, which retain a long umbilical connection to the placenta during the early stages of suckling. They detach from the teat when about seven weeks old and become independent when about eight weeks old. Sexual maturity is reached at four to five months.

41

Eastern Barred Bandicoot *Perameles gunnii* TL 270-370mm T 70-110mm

J.E. Wapstra

Now restricted on the mainland to a relict population in Victoria, this bandicoot remains abundant in Tasmania. Smaller than the Long-nosed Bandicoot, it is grey-brown above and slate-grey below, with three to four prominent pale bars on each side of the rump and greyish under parts; the tail is white. During the day it sleeps in a domed nest of grass over a scrape in the soil or in an abandoned rabbit burrow. At night it forages – in typical bandicoot fashion – for burrowing insects, insect larvae, worms, and the succulent parts of plants. It also eats berries. Obtaining sufficient water from its food, it does not need to drink. The male and female are territorial, solitary and aggressive. Mating can occur at any time of the year. Gestation is completed in 12.5 days. The female has eight teats in a backwardly opening pouch but seldom carries more than three young. These vacate the teats when seven to eight weeks old and are suckled in the nest until about nine weeks old. The female becomes sexually mature when three to four months old. The Eastern Barred Bandicoot is preyed upon by the Fox and the feral Cat.

Western Barred Bandicoot *Perameles bougainville* HB 200-300mm T 80-120mm

Babs & Bert Wells

Once distributed over much of western and southern Australia, this bandicoot now survives only on Bernier and Dorre islands, off Shark Bay. It resembles the Eastern Barred Bandicoot but is smaller, darker (brownish-grey), has white under parts and dark, less distinct, bars on the rump. During the day it sleeps in a cup-shaped nest covered by sticks and leaves. At night it digs for burrowing insects and succulent tubers or bulbs. Mating occurs from April to October. The female has eight teats in a backwardly opening pouch but it is usual for only two to be occupied.

Rufous Spiny Bandicoot *Echymipera rufescens* HB 30-41mm T eight-10mm

G.C. Richards

This is the only Australian representative of the family Peroryctidae. It is distinguished by spiny fur and the presence of only four pairs of upper incisors. It has a markedly tapering head and tawny reddish flecks in blackish fur on the back and rump: the short, black tail is almost naked. It is nocturnal and inhabits rainforest. Insects and succulent plant material are eaten. The female has six teats and usually rears one or two young.

43

Marsupial Mole *Notoryctes typhlops* HB 120-160mm T 20-30mm

This small, elusive species is placed in an order of its own, the Notoryctemorpha. By far the most specialised of all marsupials, it lacks eyes and external ears; the snout is covered by a horny shield; the tail is stubby and hairless; the fur covering the head and body is long, silky and variable in colour from cream to golden-red. Widely distributed through the central deserts in sand dunes, flat areas between dunes and in sandy river flats, it is entirely subterranean, 'swimming' through the sand (usually about 10 to 20cm below the surface but sometimes at a depth of more than 2m) with its short, powerful limbs, the claws of which are spade-like. It does not make a burrow, displaced sand falling in behind it as it moves forward. It seldom comes to the surface except after heavy rains which may water-log the soil. Fragmentary information indicates that it feeds on burrowing insects and their larvae, centipedes and lizards. The male lacks a scrotum, the testes lying within the body wall. The females has a well-developed, backwardly opening pouch enclosing two teats: the male also has a pouch. Nothing is known of its reproduction. A second, very similar species, *Notoryctes caurinus*, described in 1920, has been disregarded by most mammalogists but current research indicates that it may be a reality.

Species of *Notoryctes* not dealt with: *N. caurinus.*

Common Wombat *Vombatus ursinus* HB 900-1200, T 20-30mm

G. Hoye

The family Vombatidae, which contains three species in two genera, is a relic of a much larger group, most members of which became extinct before humans reached Australia. The Wombat has a large, broad skull and have only one pair of incisors in the upper and lower jaws; the limbs are short and powerfully clawed; the tail is a stub. The Common Wombat, only member of the genus *Vombatus*, has coarse hair and a naked muzzle. It inhabits wet and dry sclerophyll forest with adjacent grassland into which it moves at night to graze. During the day it sleeps in a long and often complex burrow it excavates with four limbs; this is usually 2 to 5m long but may extend to 20m. An individual may move between 12 or so burrows, visiting up to four in a night. The male and female are territorial and aggressive and may occupy overlapping home ranges, but usually avoid combat. Individual feeding areas are marked by scent and dung deposits. Mating can occur throughout the year with a peak in spring. The female has two teats in a large, backwardly directed pouch but rarely rears more than one young, which remains in the pouch for about eight to nine months but may accompany the mother at foot until completely weaned when about 12 months. Few individuals become sexually mature before the age of three years.

45

Southern Hairy-nosed Wombat *Lasiorhinus latifrons* HB 770-1 000mm
T 20-60mm

H. & J. Beste

The two species of *Lasiorhinus* are characterized by a blunt head, hairy muzzle and fine
silky body hair. The Northern Hairy-nosed
Wombat is on the verge of extinction; the
southern species, from the eastern Nullarbor
Plain, may be vulnerable. Externally, these
species are very similar, but readily separated in terms of distribution. The Southern Hairy-nosed Wombat is well adapted to arid
conditions, spending the day in a deep burrow in a communal warren. Each warren is home to five to ten individuals and the warrens
themselves are frequently in groups, leading to high local densi-
ties. When in a burrow, a wombat's temperature drops considerably, thus conserving energy and reducing the loss of water by
evaporation. At night it emerges to feed on grasses and herbs. Mating tends to occur from August to November but may be delayed
for several years at a time if the sporadic rainfall in its habitat is
insufficient to bring on a flush of plant growth. The female has two
teats in a large, backwardly directed pouch but it is usual for only
one young to be reared, remaining in the pouch until about nine
months old and thereafter following its mother at heel until becoming independent when about 12 months. The female become sexually mature when about three years old. Under unfavourable conditions the rate of reproduction is very low.

Species of *Lasiorhinus* not dealt with: *L. krefftii.*

Koala *Phascolarctos cinereus* HB 500-820mm T 10mm

Only member of the family Phascolarctidae, the Koala has an extensive but disjunct distribution from northern Queensland to southern Victoria. The Koala is characterized by a rounded head, dense woolly fur, long limbs with strong claws, and an extremely short tail. In a cline from north to south, the body size increases, ears become shorter, body size increases and fur colour changes from light grey to brownish-grey. Essentially arboreal and an extremely agile climber and leaper, it can gallop swiftly on the ground and is an excellent swimmer. It is restricted to eucalypt forest and feeds almost exclusively on the leaves of eucalypts, varying from place to place in its choice of food species. It is nocturnal and crepuscular, but may be active by day in cool weather. It does not use a den or make a shelter. The male is larger than the female, aggressive and territorial. The female has two teats in a backwardly opening pouch but rarely rears more than a single young. After detaching from the teat at the age of five to six months, it is carried on the mother's back until becoming independent when about 12 months old. The male disperses in search of territory; the female tends to remain in the area of birth. Although its distribution has decreased considerably since European settlement, it remains a relatively abundant species.

Mountain Pygmy-possum *Burramys parvus* HB 100-120mm T 140-150mm

G. Hoye

Although this species gives its name to the family, Burramyidae, it is an atypical member. Whereas other pygmy-possums are essentially arboreal, the Mountain Pygmy-possum lives mainly on the ground, spending the depth of the winter under snow. Distinguished by its behaviour and alpine habitat, it also differs from the other pygmy-possums in being the largest, having relatively smaller eyes, a short muzzle and a narrow tail that is never swollen at the base. During the day it sleeps in a nest of dry grass under vegetation and close to the ground. In cold conditions, or when food is scarce, it becomes torpid for days at a time. At night (or by day, in the coldest part of the year) it forages for terrestrial arthropods. In spring and summer it also collects seeds, which are stored near the nest for consumption in winter. The male is a little larger than the female, but it is not particularly aggressive. Mating occurs in October and November. The female has four teats in a forwardly opening pouch and usually carries four young, which detach from the teat when about five weeks old and are suckled in the nest until seven to eight weeks old. The young become independent when 10 to 11 weeks old, but young females may share a nest with their mother for 12 months or more before becoming independent when two years old.

Long-tailed Pygmy-possum *Cercartetus caudatus* HB 100-110mm T 130-150mm

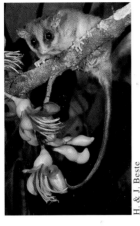

Species of *Cercartetus* are characterized by a stocky body, relatively pointed snout, short rounded ears, large eyes and a long, naked prehensile tail that is broad at the base. The tongue is long and brush-tipped. The Long-tailed Pygmy-possum is the largest of the four species and the only one in northern Queensland. It sleeps by day in a spherical nest of leaves in a tree-hole, often becoming torpid. It is arboreal and feeds on insects and nectar. The long, thin tail is used in climbing and in carrying nesting material. The female has a forwardly opening pouch enclosing four teats and may carry up to four young, which become detached when six to seven weeks old and are weaned when nine to 11 weeks old. Juveniles may share a nest with their mother.

H. & J. Beste

Species of *Cercartetus* not dealt with:
C. lepidus.

Western Pygmy-possum *Cercartetus concinnus* HB 70-110mm T 70-96mm

Fawn to reddish-brown above and white below, this pygmy-possum has a tail that is naked over most of its length. It is dependent on low, dense, profusely flowering shrubs, particularly banksias, where it feeds at night on insects and nectar. It is an agile climber and makes long leaps; it also forages on the ground. By day it sleeps in a leaf-lined nest in a tree-hole or on the ground, usually becoming torpid. Breeding occurs at any time and up to three litters may be reared in a year. The female has a forwardly opening pouch with six teats and carries up to six young for four to five weeks. They are suckled in a nest until about seven weeks old.

J. Lochman

Eastern Pygmy-possum *Cercartetus nanus* HB 70-110mm T 75-105mm

K. Atkinson

Restricted to well-watered forests, woodland and woodland and heathland in the south-eastern mainland and Tasmania, this pygmy-possum is fawn above, pale below, and distinguished by an almost naked, usually carrot-shaped tail, about the same length as the body. In Tasmania its range overlaps with that of the Little Pygmy-possum, from which it is distinguished by larger size and white-tipped hairs on the under parts. During the day it sleeps in one of several nests of shredded bark, usually located in a tree-hole, but sometimes in an abandoned bird nest. Particularly during the winter, it becomes torpid for many days at a time. At night it feeds largely on nectar and pollen, supplemented by fruits and insects such as beetles, termites and moths. When food is scarce, it resorbs fat from the tail, which becomes less carrot-shaped. Breeding occurs on the mainland from September to March; in Tasmania from August to October. Two litters (rarely three) may be reared in a year. The female has a shallow pouch with up to six teats but usually only four of these are functional. Up to four young are carried on the teats for four to five weeks and are thereafter suckled in a nest until eight to nine weeks old. The female can become sexually mature when four to five months old.

50

Common Spotted Cuscus *Spilocuscus maculatus* HB 350-580mm T 310-550mm
(Until recently known as *Phalanger maculatus*)

L. Leung

This cuscus has a round head, short ears, forwardly directed eyes, woolly fur and a prehensile tail, much of the distal end of which is bare. In Australia, the male Common Spotted Cuscus has grey fur with whitish spots; the female is usually unspotted. It spends the day seated on a branch, head tucked between the knees. At night it feeds in the rainforest canopy, mostly on leaves, also on flowers and insects. It is solitary and the male is aggressive. Breeding occurs at any time of year. The female has four teats in a forwardly opening pouch but rarely rears more than one young. After quitting the pouch, the young is carried for some weeks on the mother's back.

Southern Common Cuscus *Phalanger intercastellanus* HB 350-400mm
T 280-350mm
(Until recently known as *Phalanger orientalis* or Grey Cuscus)

Also occurring in the islands of eastern New Guinea, this differs from the Common Spotted Cuscus in being greyish-brown above, with a dark stripe from between the eyes to the rump, longer snout and ears, and an absence of spots. By day it sleeps in a tree-hole or the base of an epiphyte. At night it feeds on fruits, seeds, buds and leaves of rainforest trees. Individuals are solitary. The time of breeding is unknown. The female has four teats in a forwardly opening pouch and usually rears two young, which ride on the mother's back until weaned.

C.A. Henley

Common Brush-tail Possum *Trichosurus vulpecula* HB 350-550mm
T 250-400mmm.

Dick Whitford

The two species of *Trichosurus*, restricted to Australia, are stoutly built marsupials with a rather short, blunt muzzle, relatively large ears and eyes, and a well-furred, moderately prehensile tail which has a narrow naked area on the underside for much of its distal length. Unlike other possums, it is unable to make a grip between the first two and other three digits of the hand. The Common Brush-tail differs from the Mountain Brush-tail in having larger ears and, in northern Queensland, there is a copper-coloured subspecies. There is a marked cline in size,

individuals in the southern part of the range being much larger than those in the north. It sleeps by day in a nest in a tree-hole, under a rock or fallen timber, or in the roof-spaces of houses. Although feeding mostly on the leaves of eucalypts, it also eats fruits, buds and the softer grasses. It is solitary, territorial and aggressive. It breeds from March to May. The female has two teats in a well-developed, forwardly opening pouch, but usually rears only one young, which remains in the pouch for four to five months and is suckled until up to seven months old. In the later stages of weaning, it rides on its mother's back. In recent years the north-western population of the Common Brush-tail was regarded as a separate species, *Trichosurus arnhemensis*. The current view is that this is a subspecies, *T. v. arnhemensis*.

Mountain Brush-tail Possum *Trichosurus caninus* HB 400-500mm
T 340-420mm

H. & J. Beste

Sometimes known in Victoria as the 'Bobuck', this species is a little larger than the Common Brush-tail Possum and differs from it in having darker fur, a more stocky body, bushier tail, and shorter, rounder ears. A black form occurs in north-eastern New South Wales. Despite its common name, this brush-tail possum does not prefer montane regions but, since most of its original habitat on the coastal plains of south-eastern Australia has been cleared, it is now most abundant along the Great Dividing Range. During the day it sleeps in a den in a tree-hole, hollow log, or among epiphytes. At night it feeds mostly in trees or on leaves, fruits, buds, fungi and lichens, but also descends to the ground in search of these items; at such times it often falls prey to the Dingo. The males has a scent gland on the chest but, since its secretion is colourless, this does not lead (as in the Common Brush-tail) to staining of the pale fur of the underparts. Individuals are solitary, territorial and aggressive. Breeding occurs from March to May. The female has a well-developed, forwardly opening pouch with two teats, but usually only one young is reared. It remains in the pouch for five to six months and is suckled until about nine to 11 months old. In the later months of weaning it rides on its mother's back.

Scaly-tailed Possum *Wyulda squamicaudata* HB 310-395mm T 255-350mm

This genus, which has only one species, is restricted to forests and vine thickets in rocky areas of the Kimberley region. It is grey above and has a longer, more tapering snout than a brushtail possum. The ears are short and the very prehensile tail is covered with small tubercles, except at the base, which is well furred. Little is known of its biology. It sleeps during the day in the shelter of a rock-pile, emerging at night to feed in trees. The full range of its diet has not yet been determined but it has been seen to eat flowers. In captivity it eats leaves, fruits, nuts and insects. In many respects the Scaly-tailed Possum is intermediate between brushtail possums and cuscuses, but its combination of ground-nesting and arboreal feeding is unusual (although shared with the Rock Ringtail Possum, *Pseudocheirus dahli*). The species is very patchily distributed throughout its range and its habitat requirements have not yet been determined; it may be restricted to very rugged sandstone country. Little is known of breeding, but births tend to occur in the dry season (March – August). The female has a well-developed, forward-ly opening pouch enclosing two teats, but usually only one young is reared. It vacates the pouch when 21 to 27 weeks old and is weaned when about eight months old.

Babs & Bert Wells

Common Striped Possum *Dactylopsila trivirgata* HB 240-280mm
T 315-390mm

Dick Whitford

This possum is boldly striped in black and white: the underparts are pale. The fourth digit of the forepaw is considerably longer than the others. It has a strong, musty odour. It spends the day in a nest of leaves in a tree-hole. At night it moves rapidly and noisily through the canopy, often leaping between branches. It feeds mainly on grubs that bore into branches, biting into the timber with its incisors and extract-

ing the insect with its tongue or the claw of its fourth finger. Breeding is from February to August. The female has two teats in a forwardly opening pouch and rears one or two young.

Leadbeater's Possum *Gymnobelideus leadbeateri* HB 150-170mm
T 150-180mm

A. Smith

This species resembles *Petaurus* but lacks a gliding membrane. During the day it sleeps in a nest of shredded bark in a hole in an old Mountain Ash tree. At night it feeds in the canopy on gums of acacias and eucalypts, manna and medium-sized arthropods. A pair shares a nest with up to six offspring, including non-breeding males which jointly defend a territory. Mating is from April to May or September to October. The female has four teats in a forwardly opening pouch and usually carries one or two young that leave the pouch when 11 to 13 weeks old and are suckled in the nest until about 16 weeks old.

55

Sugar Glider *Petaurus breviceps* HB 160-210mm T 170-210mm

R.W. Jenkins

The three species of *Petaurus* are possums with a gliding membrane extending from the fifth digit of the forepaws to the ankles; the tail is well furred, restricted at its junction with the body, and weakly prehensile; there is a distinct, dark, mid-dorsal stripe. The Sugar Glider resembles the Squirrel Glider but is slightly smaller, has a blunter snout, denser body fur and a narrower base to the tail. By day it rests in a leaf-lined nest in a tree-hole with up to six other adults and their young. When cold or when food is scarce, it may become torpid. Sharing a common odour, a family defends a scent-marked territory from other groups. In winter it feeds mainly on gummy exudates of eucalypts and acacias; at other times, insects predominate. The Sugar Glider encourages exudates by cutting grooves in the trunk with the incisors. Family groups defend such grooves and keep them flowing. In open forest it moves from one tall tree to another by gliding over distances of up to 50m. Breeding occurs from June to November and two litters may be reared in a year. The female has a forwardly opening pouch enclosing four teats and often rears two young. These remain in the pouch until about 10 weeks old and are suckled in a nest until about 15 weeks old. Independence is reached at the age of about 17 weeks but off-spring may remain much longer in the parental nest.

Squirrel Glider *Petaurus norfolcensis* HB 170-240mm T 220-300mm

Dick Whitford

This resembles the Sugar Glider and can interbreed with it, but it is larger (about twice the weight), has a somewhat longer snout, fluffier fur, broader-based and a distinctly tapering tail. The fur on the under parts is white. Its habitats range from dry sclerophyll forest and woodland in south-eastern Australia to wet sclerophyll in northern New South Wales and Queensland. It sleeps by day in a leaf-lined nest in a tree-hole. Typically a nest includes a male, one or more females, and their sexually immature young; as many as five adults and five young have been found in a nest. At night the Squir-

rel Glider moves rapidly through the forest, feeding on insects, nectar and pollen, and the gummy exudates of eucalypts and acacias, which are encouraged, as in the Sugar Glider and Yellow-bellied Glider, by incising grooves in the trunks of trees. To move across a gap in the forest an individual climbs a tree (often a tall, dead one), extends the four limbs and glides to the base of another tree. Breeding occurs from May to December. The female has four teats in a well-developed, forwardly opening pouch and usually carries two young. These leave the pouch when about 10 weeks old and are suckled in the nest until about 15 weeks old. Young accompany their mother until about 17 weeks old, after which they may disperse or remain in the parental nest.

Yellow-bellied Glider *Petaurus australis* HB 270-300mm T 420-480mm

More than twice the weight of a Sugar Glider, this is the largest species of *Petaurus* and has a proportionately longer tail (1.5 times the head-body length). The underparts are cream to orange; the ears are long and almost naked. It inhabits tall eucalypt forest and woodland and spends the day in a leaf-lined nest in a tree-hole. Nests are usually shared. In the cooler part of the range, an adult male may associate with an adult female and her young; in the warmer part, a male may share a nest with two adult females and up to three young. The male is territorial and marks its territory with scent from glands on the head, chest and tail. At night it moves through the canopy, feeding on nectar, pollen and the exudates of insects and eucalypts. It makes and maintains incisions in the trunks of smooth-barked eucalypts to promote the flow of sap and gum. Insects and spiders are also eaten. It frequently utters loud, high-pitched screams and throaty gurgles. Breeding occurs in spring in the southern part of the range but throughout the year (with a peak in July and August) in the northern part. The female has a forwardly opening pouch that is partially divided into right and left compartments, each with a teat. The single young remains in the pouch for 13 to 14 weeks and is thereafter suckled in the nest for six to nine weeks.

Lemuroid Ringtail Possum *Hemibelideus lemuroides* HB 310-350mm T. 33-37mm

H. & J. Beste

A feature of this ringtail, is its flattish 'lemur-like' face. The weakly prehensile tail is bushy, with little taper from the base, and has a short bare tip. It is grey, without markedly pale underparts. It makes long leaps from one branch to another. By day it sleeps in a nest in a tree-hole, often with a mate and young. At night it feeds mainly on leaves of rainforest trees. Breeding is mainly from July to October. The female has two teats in a forwardly directed and divided pouch but normally rears only one young, which rides on its mother's back after leaving the pouch.

Greater Glider *Petauroides volans* HB 350-450mm T 450-600mm

C.A. Henley

Largest of the gliding marsupials, this species is distinguished by a gliding membrane from the elbows (not the wrists) to the ankles: when gliding, the forearm is flexed at the elbow, with the forepaws under the chin. The tail is long, well furred, untapered and weakly prehensile. Upperparts vary from grey to mottled or chocolate brown. By day it sleeps in a nest in a tree-hole. A pair usually share a nest from the onset of mating until young leave the pouch. The Greater Glider feeds almost exclusively on eucalypt leaves. Breeding is from March to June. The female has two teats in a forwardly opening pouch but usually rears one young, which leaves the pouch at 12 to 16 weeks and is suckled in the nest until able to ride on its mother's back.

Common Ringtail Possum *Pseudocheirus peregrinus* HB 300-390mm
T 300-350mm

G.B. Baker

The genus *Pseudocheirus* includes two species, characterized by a tapering tail which is usually short-furred and with a bare area under the terminal part. The Common Ringtail Possum is greyish above with a rufous tinge, particularly on the limbs, and has white patches behind the eyes and under the ears, which are short and rounded. The tip of the very prehensile tail is white. It occurs from rainforest to woodland over most of the better-watered areas of eastern Australia and south-western Australia, often in suburbs. During the day it sleeps in one of several well-constructed spherical nests (dreys) made of grass and shredded bark and built in a tree-hole or fork. Several individuals may share a nest. At night it moves rather deliberately through the foliage, using the tail and a 'fifth limb'. It feeds mainly on the leaves of eucalypts, but also eats flowers, fruits and buds. Individuals are territorial and mark their territories with scent secretions from glands near the vent. Breeding occurs from April to November. The female has four teats in a forwardly opening pouch and suckles two young. Two litters may be reared in a year. Young leave the pouch when six to seven weeks old and are suckled for a time in a nest. Later they ride on the mother's back or follow at heel. They are not weaned until six months old.

Species of *Pseudocheirus* not dealt with: *P. occidentalis.*

Herbert River Ringtail Possum *Pseudochirulus herbertensis* HB 300-375mm T 320 -400mm

Very similar to *Pseudocheirus*, this genus has two species in Australia. The tapered snout, small ears and forwardly directed eyes distinguish this rainforest species from other Australian ringtails. There is a dark stripe between the eyes. The strongly prehensile tail, which is bare for most of its underside, tapers to a narrow tip, often white. By day it sleeps in a tree-hole or a spherical nest. At night it feeds on leaves of rainforest trees. Mating is mainly from April to May. The female has two teats in a forwardly opening pouch and usually rears two young, which quit the pouch at 16 to 20 weeks, then ride on the mother's back for a week or two before being suckled in the nest until 21 to 23 weeks old.

H. & J. Beste

Species of *Pseudochirulus* not dealt with: *P. occidentalis.*

Green Ringtail Possum *Pseudochirops archeri* HB 340- 380mm T 310-330mm

Pseudocheirops is distinguished from *Pseudocheirus* by larger size, larger and more complex molar teeth, and a complex pigmentation of the hairs that confers an unusual copper, silver or green colouration to the fur. The Green Ringtail is large with greenish, woolly fur, a thick-based, well-furred, tapering tail, and prominent white patches under the eyes and ears. By day it sits on a branch with head between its knees, the prehensile tail is coiled. At night it feeds on the leaves of rainforest trees. It does not leap. Breeding appears to peak from July to September. The female has two teats in a forwardly opening pouch and usually rears one young, which rides on the mother's back after leaving the pouch.

H. & J. Beste

Feathertail Glider *Acrobates pygmaeus* HB 60-80mm T 70-80mm

P. German

This small marsupial, only member of its genus, is readily distinguished from all other marsupials by the feather-like arrangement of stiff, closely packed hairs on either side of the weakly prehensile tail, and possession of a gliding membrane that extends from the elbows to the knees. It is widely distributed in eastern Australia from sclerophyll forest to woodland. During the day it sleeps, often communally, in a well-constructed spherical nest of leaves in a tree-hole or a wide variety of other niches. In cold weather it may become torpid. As many as 16 individuals have been found in one nest (22 in captivity). At night it moves rapidly through the canopy, sometimes making leaps of up to 30m. The fingers and toes have large, finely serrated pads under their tips, providing adhesion to smooth surfaces. It feeds on nectar, pollen and small insects. Breeding can occur at any time of the year in the northern part of the range, but not in autumn or early winter in the south. The female has four teats in a forwardly openly pouch and often rears three young; several litters may be reared in a year. Young leave the pouch when about nine weeks old and are suckled in a nest until becoming independent at the age of about 14 weeks. Until recently *Acrobates* was regarded as a pygmy-possum of the family Burramyidae. It is now placed in the family Acrobatidae, with one other member, the much larger, non-gliding Feathertail Possum, from New Guinea.

Honey-possum *Tarsipes rostratus* HB 40-95mm T 45-110mm

Babs & Bert Wells

The only member of its genus and family, this small marsupial is characterized by a long, pointed snout; and eyes that appear to be upwardly directed. There is a dark stripe along the middle of the back, flanked by a pale stripe and then a slightly darker one. The tail is long, lightly furred and gradually tapers to a very fine tip; it is very prehensile. The claws on the fingers and toes are nail-like except on the conjoined second and third toes of the hindfoot, which are sharp and narrow. During the day it sleeps in an abandoned bird nest or the hollow stem of a grass-tree, often becoming torpid. At night (or during the day, in cold weather) it moves rapidly through low scrub or heath in search of blossoms from which it imbibes nectar and pollen with its long, brush-tipped tongue. Pollen adheres to its head and body when it feeds and its contribution to pollinating many plants, particularly banksias, is significant. Breeding occurs throughout the year, but less often in summer (when nectar is least abundant). The female has four teats in a well-developed, forwardly opening pouch and usually rears two or three young, which leave the pouch when about eight weeks old and are suckled in a nest or shelter until about 10 weeks old. The Honey-possum is only distantly related to the other possums; recent research indicates that it has some affinity with the Feather-tail Glider.

Musky Rat-kangaroo *Hypsiprymnodon moschatus* HB 210-275mm T 125-160mm

Dick Whitford

Kangaroos undoubtedly evolved from possum-like ancestors and an indication of what the intermediate forms may have been like is provided by this species, the only member of its genus, which is distinguished from all other kangaroo-like marsupials by having five toes on the hindfoot (the others lack the first digit or 'big toe') and two pairs of lower incisors (one pair very small). The difference in size of the forelimbs and hindlimbs is much less than in kangaroos. It is very unusual among marsupials in being predominantly diurnal, spending the night in a nest of leaves, often made between the buttress roots of a large rainforest tree. Although mainly terrestrial, it climbs in low vegetation, using the first digit of the hindfoot to grip branches. It does not hop, but moves over the ground in a quadrupedal walk, or bounds like a Rabbit. It is solitary and omnivorous, eating insects, fallen fruits and nuts and fungi. Food is held in the forepaws while being eaten. The seeds of some fruits are not eaten immediately, but buried in the leaf-litter for later consumption. Mating occurs between April and October. The female has four teats in a forwardly opening pouch and usually rears two young, which leave the pouch when about 21 weeks old and are suckled in a nest for several weeks.

Brush-tailed Bettong *Bettongia penicillata* HB 300-380mm T 290-360mm

Babs & Bert Wells

Three species are placed in this genus of rat-kangaroos, characterized by a short, broad head and (in comparison with potoroos) a plumper body, longer tail (about as long as the head and body) and a hindfoot that is longer than the head. The Brush-tailed Bettong is distinguished by a black, crested brush on the distal-third of the tail. Once widespread over much of the woodland and tussock grassland of southern Australia, it is now restricted to small areas in Western Australia and Queensland. It is solitary and aggressively territorial. During the day it sleeps in a well-made, dome-shaped nest situated over a scrape in the ground, usually under a bush. Grass and shredded bark is carried to the nest in the prehensile tail. At night it forages for bulbs, tubers and underground fungi, excavating these with its powerful, strongly clawed forepaws. It also eats insects and the resin of *Hakea* shrubs. Breeding occurs throughout the year. The female has four teats in a forwardly opening pouch but rarely carries more than one young, which quits the pouch when about 13 weeks old and follows its mother at heel, sharing her nest until about 17 weeks old. The female becomes sexually mature when about 24 weeks old and, under favourable conditions, produces one young every 14 weeks.

Species of *Bettongia* not dealt with: *B. lesueur*

Tasmanian Bettong *Bettongia gaimardi* HB 315-330mm T 290-345mm

D. Watts

This is the only bettong in Tasmania. The prehensile tail has a white tip. Solitary and territorial, it lives in dry sclerophyll forest with a grassy understorey. By day it sleeps in one of several nests of woven grass and bark, usually in a shallow excavation. At night it digs for underground fungi, roots and bulbs; it seems not to eat any green food. It is solitary and territorial, occupying a large home range. The female has four teats in a forwardly opening pouch and rears one young, which quits the pouch at about 15 weeks and follows its mother until about 22 weeks old. The female becomes sexually mature at about 12 months and can produce two or three young in a year.

Rufous Bettong *Aepyprymus rufescens* HB 375-390mm T 340-390mm

G. Hoye

Closely related to *Bettongia*, this genus is distinguished by the presence of hairs on the central part of the muzzle. The head is broad, the ears pointed and the fur bristly. Largest of the bettongs, it inhabits well-grassed open forest. By day it sleeps in a nest of vegetation constructed over a shallow scrape, usually under a tussock or fallen log. Each individual has a number of nests in its territory. It is nocturnal and aggressive. It eats grasses, sedges and herbs and also digs up tubers with its strongly clawed forepaws. The prehensile tail is used to carry nesting material. Breeding is continuous. The female has four teats in a forwardly opening pouch, and rears one young which leaves the pouch at about 16 weeks old and follows its mother until about 23 weeks old.

Long-nosed Potoroo *Potorous tridactylus* HB 320-400mm T 200-260mm

J. Fennell

The genus *Potorous* is distinguishable from other rat-kangaroos (*Bettongia*, *Aepyprymnus*) by a slender, tapering head and a hindfoot that is shorter than the head. The two living species are more slender than bettongs and have a relatively shorter tail. The rhinarium is bare and a strip of naked skin extends from it onto the upper surface of the snout. There is considerable local variation and from west to east across northern Tasmania the average weight doubles and colour of the upper parts changes from rufous brown to grey-brown. All have very restricted distributions but the Long-nosed Potoroo is reasonably well distributed in cool rainforest, wet sclerophyll forest and heathland in the south-western coastal region of the mainland and Tasmania. It is smaller than the Long-footed Potoroo, *P. longipes*, which it otherwise resembles. By day it sleeps in a roughly constructed nest of vegetation over a scrape in the ground, usually in the shelter of a shrub or tussock. Nesting material is carried to the site in the prehensile tail. At night it uses its strongly clawed forepaws to dig for succulent roots, tubers, fungi and subterranean insects. Breeding occurs throughout the year, with peaks from late winter to early spring and in late summer. The female has a well-developed, forwardly opening pouch, but carries only one young, which remains in the pouch until about 16 weeks old, then follows its mother, sharing its nest, until about 22 weeks old. The female becomes sexually mature at the age of 12 months.

Species of *Potorous* not dealt with: *P. longipes*.

67

Bennett's Tree-kangaroo *Dendrolagus bennettianus* HB 690-750mm, T 730-840mm

L.J. Roberts

Tree-kangaroos are primarily New Guinean but two species occur in Queensland rainforest. They differ from all other kangaroos in having powerful forelimbs, relatively short, broad hindfeet, and a long, cylindrical pendulous tail. Bennett's Tree-kangaroo has a black patch at the base of the tail and a pale patch above. Solitary, territorial and aggressive, it spends the day seated on a branch with its head between its knees and its tail dangling. At night it feeds on the leaves and fruit of rainforest trees. The male is larger than the female, which has four teats and rears one young, which quits the pouch at about nine months and follows its mother until almost three years old.

Lumholtz's Tree-kangaroo *Dendrolagus lumholtzi* HB 480-590mm T 600-740mm

D. Watts

Similar to Bennett's Tree-kangaroo, this is distinguished by a pale band across the forehead and on the sides of the face. It spends the day seated on a branch, with tail dangling. A parting of the fur, just behind the shoulder, diverts rain forwards over the head and backwards over the rest of the body. It is solitary and probably territorial. At night it feeds on leaves and fruits of rainforest trees. The female has four teats in a forwardly opening pouch and rears one young which spends about 33 weeks in the pouch.

Spectacled Hare-wallaby *Lagorchestes conspicillatus* HB 400-470mm
T 370-390mm

H. & J. Beste

Species of *Lagorchestes* are related to the majority of living kangaroos (subfamily Macropodinae): the Banded Hare-wallaby, *Lagostrophus fasciatus*, (p. 83) is the only surviving member of the subfamily Sthenurinae. *Lagorchestes* has a broad head, hairy muzzle, shaggy fur, short forelimbs, and a long, moderately haired, tapering tail. The Spectacled Hare-wallaby is distinguished by pronounced reddish rings around its eyes. By day it sleeps in a den at the base of a tussock or hummock. At night it feeds on leaves of shrubs and tips of spinifex; it does not drink. The female has four teats but rears only one young, which quits the pouch at about 21 weeks.

Rufous Hare-wallaby *Lagorchestes hirsutus* HB 310-390mm T 250-300mm

Babs & Bert Wells

This compact wallaby is distinguished by a strongly rufous tinge to the upper parts, pale forearms, soft fur and slender, almost untapered tail. It is now restricted to the Tanami Desert (where it was reintroduced) and two islands off Shark

Bay. Solitary and nocturnal, it spends the day in a shallow excavation in a spinifex hummock, emerging to feed at night on the tough parts of such plants as sedges, grasses and herbs, supplemented by insects. Breeding can occur at any time of the year, but this is determined by rainfall. The female is larger than the male.

Red Kangaroo *Macropus rufus* HB 740-1400mm T 640-1000mm

G. Little

The genus *Macropus*, with 13 living species, is the most successful of the family Macropodidae. It is difficult to define the genus but members are characterized by a grazing habit, large grinding molars, and a compartmentalized stomach that permits the digestion of very fibrous grasses. The Red Kangaroo is distinguished by its large size and, in the eastern part of the range, red upper parts in the male; the female tends to be blue-grey, as are some males in the western part. The muzzle is naked and the ears long. The forearms of the mature male are large and muscular. Well adapted to arid conditions, it drinks when water is available, but is able to survive without water if sufficient succulent green feed is available. By day it rests on its side in a shallow scrape in the shade of a bush or shrub. At night it moves into open country to graze. The male, which is up to three times the weight of the female, is territorial and attempts to monopolize the reproduction of a number of females. At night the Red Kangaroo moves into open country to graze. Breeding can occur at any time of the year but the production of young tends to be in response to unpredictable rainfall. The female has four teats in a forwardly directed pouch and rears a single young, which quits the pouch at about 38 weeks, remaining dependent until nearly 12 months old. The female becomes sexually mature in its second year.

Species of *Macropus* not dealt with: *M. bernardus, M. irma.*

70

Agile Wallaby *Macropus agilis* HB 590-850mm T 590-840mm

G.B. Baker

This slender wallaby has generally sandy upper-parts and white underparts, black margins to the ears, and a distinct white hip-stripe. The tail is long and only slightly tapering. It inhabits forest adjacent to open country, emerging late in the day and at night to feed on native grasses and shrubs. Breeding occurs throughout the year. The female, which is about half the weight of the mature male, has four teats in a forwardly directed pouch and rears a single young, which quits the pouch at about 30 weeks and remains dependent until about 12 months old.

Antilopine Wallaroo *Macropus antilopinus* HB 780-1 200mm T. 680-890mm

T.S.C. Smith

This large kangaroo from northern Australian woodlands is distinguished by rather soft fur, reddish to sandy upper parts and whitish under parts. The lower jaw is pale and there is no definite cheek-stripe, the tail is thick at the base and tapers only gradually. By day it rests under a bush, often near water. At night it grazes in open country. The female, which is about half the weight of the mature male, has four teats in a forwardly opening pouch and rears one young which quits the pouch at about 39 weeks but remains dependent until about 12 months old.

Black-striped Wallaby *Macropus dorsalis* HB 1 100-1 600mm T 540-830mm

C.A. Henley

This is distinguished by brown upperparts and a distinct mid-dorsal stripe from the neck to the rump, very pale underparts, and a white spot on the cheek, behind the eye. When hopping, it usually extends its arms sideways. Widespread through the wettish parts of southern Queensland and northern New South Wales, on either side of the Great Dividing Range, it spends much of its time under cover. During the day it often rests in groups under the shelter of shrubs. At dusk it moves along established runways to open country to feed on grasses and herbs. The female has four teats in a forwardly opening pouch but rears a single young in the pouch until about 30 weeks old. The female becomes sexually mature in its second year.

Tammar Wallaby *Macropus eugenii* HB 520-680mm T 330-450mm

Babs & Bert Wells

This small wallaby has grizzled, grey-brown upper parts, is pale grey below, slightly rufous on the sides, with a white cheek-stripe and a dark stripe between the ears. The tail is rather short. The forelimbs are extended when hopping. It spends the day in the shelter of dense vegetation and grazes at night in open areas. Adapted to aridity it is able occasionally to drink salt water. Breeding is mainly from summer to early winter and most young are born in December. The female has four teats in a forwardly opening pouch and rears a single young in the pouch for 36 weeks. The female becomes sexually mature at 32 weeks, while still suckling.

Western Grey Kangaroo *Macropus fuliginosus* HB 950-2 230mm T 425-1 000mm

Babs & Bert Wells

This large kangaroo has a finely haired muzzle and brownish upper parts. It occupies slightly drier habitats than the Eastern Grey Kangaroo. By day it sleeps in the shade; it grazes at night in groups. The male is much larger than the female. Breeding occurs at any time of the year. The female has four teats in a forwardly directed pouch and rears a single young that quits the pouch at about 10 months and remains dependent until about 12 months.

Eastern Grey Kangaroo *Macropus giganteus* HB 950-2 300mm T 430-1 100mm

A. Young

This kangaroo closely resembles the Western Grey Kangaroo, but has grey fur and occupies a less arid habitat. It moves in groups of up to 50 individuals led by a mature male. It sleeps by day in the shade of vegetation and grazes at night in open country. Breeding occurs at any time of year. The female, less than half the weight of the mature male, has four teats in a forwardly opening pouch and rears a single young, which quits the pouch at about 10 months. The female becomes mature at about 16 months.

73

Parma Wallaby *Macropus parma* HB 450-530mm T 400-550mm

This small wallaby has greyish-brown upper parts with a dark mid-dorsal stripe, a poorly defined white cheek-stripe and (often) a white tip to the tail, which is as long as the head and body. When hopping, the forelimbs are held alongside the body. By day it sleeps in low vegetation. At night it feeds on grasses and herbs. Mating occurs from January to May. The much smaller female has four teats in a forwardly directed pouch and rears one young, which quits the pouch at 36 weeks, but remains dependent until about 15 months old. The female becomes sexually mature when about 18 months old.

Babs & Bert Wells

Whiptail Wallaby *Macropus parryi* HB 675-1 000mm T 720-1 050mm

G.B. Baker

This medium-sized kangaroo has a slender, dark-tipped tail that is a little longer than the head and body; there is a prominent white cheek-stripe, light brown shoulder-stripe, and white hip-stripe. It moves in groups of up to 50 led by a male. By day it rests in low under-growth in wet or dry sclerophyll forest. At night it grazes in open country. Breeding is continuous. The female, about half the weight of the mature male, has four teats in a forwardly opening pouch and rears one young, which quits the pouch at about 40 weeks but remains dependent until about 14 months old.

Common Wallaroo/Euro *Macropus robustus* HB 1 100-2 000mm T 530-900mm

G.B. Baker

This large kangaroo extends from the eastern to the western coasts of the mainland and has evolved into four subspecies. The eastern population, on both sides of the Great Dividing Range, has shaggy grey fur and is known as the Common Wallaroo. Westward from there to the coast of Western Australia, it is shorter haired, reddish and is known as the Euro. (The other subspecies have much more limited distribution.) It occurs in a wide range of habitats from wet sclerophyll forests to arid tussock grasslands but is usually associated with rocky slopes that have caves or rock-shelves where it rests during the day. At night it emerges to graze in more level country. The Euro is arid-adapted and does not need to drink. Individuals may be solitary or form small groups. Mating can occur at any time of the year, but in arid areas may be dependent upon unpredictable rainfall. The female is about half the weight of a mature male. It has four teats in a forwardly opening pouch and rears a single young, which leaves the pouch when about 36 weeks old but remains dependent until the age of about 16 months. The female becomes mature in its second year.

Red-necked Wallaby *Macropus rufogriseus* HB 660-890mm T 620-880mm

G.B. Baker

Common over most of the forested areas of the south-eastern mainland and Tasmania, this large wallaby has grey to reddish upper parts, particularly over the neck and shoulders and is pale below; the muzzle and paws are black; there is a dark brown stripe between the eyes; the ears are black-edged and the tapering tail often has a dark tip. By day it sleeps in dense vegetation, feeding after dusk on grasses and herbs. Breeding is continuous. The much smaller female has four teats in a forwardly directed pouch and produces one young, which quits the pouch at 40 weeks but remains dependent until 12 to 17 months.

Northern Nailtail Wallaby *Onychogalea unguifera* HB 490-690mm T 600-730mm

P. Fell

Species of *Onychogalea* have a thick horny fingernail-like plate on the tip of the tail, which has a slight crest of hairs. The ears are long and pointed; there is a prominent white stripe on the cheek; the fur is fine and silky. Of three species, one is extinct and one is endangered. The Northern Nailtail Wallaby has a black tip (sometimes barred) on the terminal quarter of the tail and a prominent pale hip-stripe. It ranges from open woodland to grassland. By day it rests in a scrape in the soil under a shrub. At night it feeds on foliage and fruits. The female, which is smaller, has four teats in a forwardly opening pouch and rears one young.

Bridled Nailtail Wallaby *Onychogalea fraenata* HB 430-700mm T 360-540mm

H.J. Beste

This is distinguished by a prominent pale shoulder-stripe ('bridle'), a prominent pale cheek-stripe and a tail that is not black at the tip. It once ranged widely but is now found in open eucalypt and brigalow forest adjacent to grassy areas in a small area in central Queensland. By day it sleeps in a shallow scrape in the soil. At night it feeds in open country on leaves of forbs, herbs and some grasses. The long-clawed forepaws scrape away dry litter to expose young shoots. Breeding is continuous with a peak in late winter and early spring.

Brush-tailed Rock-wallaby *Petrogale penicillata* HB 450-580mm
T 520-670mm

Babs & Bert Wells

Rock-wallabies are extremely agile inhabitants of rocky slopes and rock-piles, leaping with the aid of a very mobile, balancing tail. Because they live on rocky slopes, they occur in many separate areas of minor elevation. This has led to great genetic isolation in what is still a rapidly evolving group of marsupials and, in consequence, there is some disagreement among zoologists about the specific or subspecific status of particular populations (some 15 species and eight subspecies). The genus *Petrogale* is characterized by broad hindfeet with granulated soles, and a long, cylindrical tail that is longer than the head and body. The Brush-tailed Rock-wallaby, best-known because of its south-eastern distribution, is distinguished by the distinct brush at the end of the tail. Regarded as common until as recently as the 1980s, it had become endangered, at least in New South Wales, in the early 1990s. It rests by day in a rock crevice, emerging in the late afternoon and night to feed on grasses, leaves, fruits, flowers and seeds. The male is larger than the female. Breeding is possible throughout the year. The females has four teats in a forwardly opening pouch and usually rears one young. It vacates the pouch when about 30 weeks old and thereafter follows its mother at heel. Very little is known of the biology of rock-wallabies but most populations have distinguishing marks or colouring.

Species of *Petrogale* not dealt with: *P. assimilis, P. brachyotis, P. burbidgei, P. coenensis, P. godmani, P. herberti, P. mareeba, P. persephone, P. rothschildi, P. sharmani.*

Nabarlek *Petrogale concinna* HB 290-350mm T 220-310mm

G.D. Sanson

Smallest of the rock-wallabies, this is distinguished by its small size, marbled rufous upper parts and the presence of supernumerary molars. It inhabits rock-piles. Alone among macropods, it produces a continuous series of molars which move forward as the front teeth drop out. By day it rests in a rock crevice. At night it grazes on blacksoil plains. It breeds continuously with a peak in summer. The females has four teats in a forwardly directed pouch and rears one young, which quits the pouch at about 26 weeks. Sexual maturity is reached at two years.

Unadorned Rock-wallaby *Petrogale inornata* HB 450-570mm T 430-640mm

Dick Witford

This rock-wallaby is characterized by its lack of distinctive colouration. Generally grey-brown to sandy, it may have a pale cheek-stripe (far less distinct than many other rock-wallabies). The tail darkens towards the tip and may have a short crest. It spends the day asleep in a rock crevice, emerging in the evening and night. The diet is not known. The female, which is smaller, has four teats in a forwardly directed pouch. Breeding occurs at any time. The single young quits the pouch at six to seven months old and follows the mother until at least 12 months old.

Black-footed Rock-wallaby *Petrogale lateralis* HB 450-580mm T 400-570mm

G. Hoye

Most widespread of the rock-wallabies, this includes some six subspecies or races from western Queensland to islands of Western Australia. It has a thick, woolly coat, particularly about the hindparts, a prominent pale cheek-stripe, and black paws. There is great variation between populations. Once widespread over the south-western half of Western Australia, it now exists only in isolated areas. The smaller female has four teats in a forwardly directed pouch and rears one young.

Yellow-footed Rock-wallaby *Petrogale xanthopus* HB 480-650mm
T 570-700mm

A. Young

This is fawn-grey above, white below, with a distinct white cheek-stripe, a rich brown stripe from the head to the tail, and pale hip-stripes. Forearms and forefeet are yellow to bright orange and the tail has brown and white rings. Adapted to semi-arid conditions, it lives compatibly in aggregations of up to 100 individuals. By day it rests in a rock crevice, emerging at night to feed on leaves and grasses. Breeding is continuous. The females has four teats in a forwardly opening pouch and rears one young, which quits the pouch at four to five weeks.

Quokka *Setonix brachyurus* HB 400-550mm T 250-310mm

H. & J. Beste

Only member of its genus, this species has a stocky body, rather short, broad face, short, rounded ears, and a short tail. Hairs on the feet cover the claws. Common on Rottnest Island, off Perth, it is sparse in southwestern Australia in areas with a dense understorey. Several individuals may sleep together by day in heathland shelter. Groups of up to 150 may occupy a common territory. It feeds on shrubs and fibrous grasses. Breeding occurs from January to March. The female, which is smaller, has four teats in a forwardly opening pouch and rears a single young, which quits the pouch at about 26 weeks.

Red-legged Pademelon *Thylogale stigmatica* HB 390-540mm T 300-480mm

R. & D. Keller

The three Australian species of *Thylogale* are thick-furred, compact-bodied, with relatively short hindfeet and evenly furred, tapering tails that are relatively shorter than those of typical wallabies. The Red-legged Pademelon is distinguished by a rufous tinge to the cheeks, forearms and inner and outer surfaces of the hindlimbs. By day it sleeps in dense foliage in wet forests. At other times it grazes and browses in open areas. The female, which is smaller, has four teats in a forwardly opening pouch, and rears one young which quits the pouch at about 26 to 28 weeks and is weaned at about 36 weeks.

Red-necked Pademelon *Thylogale thetis* TH 290-620mm T 270-510mm

This small wallaby has a rufous tinge to the neck and shoulders. It inhabits temperate coastal forests near grassland and shrubland, but in slightly drier areas than those inhabited by the Red-legged Pademelon. It sleeps during much of the day, otherwise grazing or browsing in more open areas. Breeding tends to be in autumn and spring in the northern part of the range, in summer in the south. The

G. Little

female, which is smaller, has four teats in a forwardly opening pouch and rears a single young, which quits the pouch at about 26 weeks.

Tasmanian Pademelon *Thylogale billardierii* HB 360-720mm T 320-480mm

D. Greig

The only pademelon found in Tasmania, this is dark brown to dark grey-brown above and buff below, with a rufous tinge to the belly. By day it shelters in dense undergrowth in forest or scrubland, emerging at night to feed on grasses, herbs and shrubs. It is solitary but several individuals may come together to feed. The male is territorial and vigorously attacks intruders. Breeding occurs throughout the year. The single young quits the pouch at about 200 days and follows its mother until about 10 months old.

Swamp Wallaby *Wallabia bicolor* HB 660-850mm T 640-860mm

G. Little

The only species in its genus, the Swamp Wallaby is distinguished by very dark, dense, flecked fur, rufous underparts and conspicuous pale cheek-stripe. When hopping, the head is kept low, and the tail held straight. It ranges from tropical rainforest to cool-temperate woodland. By day it sleeps in dense vegetation. At night it feeds on shrubs, ferns and grasses. Breeding is continuous. The female, which is smaller, has four teats in a forwardly directed pouch and rears a single young, which quits the pouch at eight to nine months, remaining dependent until 16 months old.

Banded Hare-wallaby *Lagostrophus fasciatus* HB 400-450mm T 350-400mm

Babs & Bert Wells

This anomalous wallaby has transverse dark bands on the rump, and hairs on the feet that cover the claws. Although called a 'hare-wallaby', it is unrelated to *Lagorchestes* (p.69), being the only living member of the subfamily Sthenurinae. It now survives only on Bernier and Dorre islands, off Shark Bay. Solitary and aggressive, it sleeps by day under shelter, emerging to browse and graze. It does not need to drink. Breeding occurs throughout the year. The female, which is larger than the male, has four teats but rears a single young, which quits the pouch at about 26 weeks and becomes independent at about 36 weeks.

Brush tailed Tree-rat *Conilurus penicillatus* HB 150-200mm T 180-220mm

H. & J. Beste

The only surviving member of its genus, this tree-rat has a long tail bearing a brush of dark hairs; the eyes are large and protuberant; the grizzled grey-brown body fur is smooth and shiny. It ranges from monsoon forest to pandanus scrub, is arboreal and spends the day in a tree-hole or similar crevice. At night it feeds in the canopy and on the ground on grasses, seeds and fruits. Breeding is continuous. The female has four teats and rears between one to three (usually two) young, which become independent at the age of about three weeks.

Common Water-rat *Hydromys chrysogaster* HB 230-370mm T 230-330mm

R.W. Jenkins

Centred in New Guinea, this genus has four species, one widespread, but sparse, in Australia. It has large, partially webbed hindfeet. The head is long and somewhat flattened, with small eyes and ears. The fur is long, dense, shining and waterproof; the tail is closely furred and tapering. It sleeps by day in a nest in a water-side burrow, hunting at night for crustaceans, frogs, mussels and large insects taken from the bottom of a river or lake. The female has four teats and usually rears three or four young, weaned at four weeks, but remaining with the mother until about eight weeks old.

False Water-Rat *Xeromys myoides* HB 80-120mm T 70-90mm

H. & J. Beste

This rodent has short, very silky hair, slate-grey to brown above, white below. The ears are short and rounded, the eyes are small. Restricted to the vicinity of well-vegetated edges of bodies of fresh water or salt water, in the tropics and subtropics, it builds a mound of soil or mud that protrudes above the water surface and contains branching tunnels and a nest, occupied by a number of individuals. At night it feeds on crustaceans on mud-flats and in shallow pools.

Forrest's Short-tailed Mouse *Leggadina forresti* HB 65-100mm T 50-75mm

Babs & Bert Wells

The genus *Leggadina* has two species, similar in many respects to *Pseudomys*, but with a short tail (less than 70% of the head and body length). Forrest's Short-tailed Mouse is grey-brown above, pale below, with long shiny fur. It occurs in the Central Desert and Pilbara. By day it sleeps in a grass-lined nest in a burrow, up to 40cm long. It feeds at night on seeds and green vegetation. It does not need to drink. Breeding follows irregular rainfall. The female has four teats and usually rears three or four young, which become independent at about four weeks.

Species of *Leggadina* not dealt with: *L. lakedownensis*.

Greater Stick-nest Rat *Leporillus conditor* HB 170-260mm T 145-180mm

H. & J. Beste

The only surviving member of this genus is restricted to the Nuyts Archipelago. It builds large nesting structures of branches and sticks over many generations. Relatively large, it has a short, blunt head, long ears, thick, soft fur, and a long thin, moderately furred tail. By day it shelters within the stick-nest, with up to 20 other individuals. At night it forages on grasses and succulent herbs. Breeding is from March to June. The female has four teats and usually rears two young, which bite onto its teats and are dragged behind it until independent at about four weeks.

Broad-toothed Rat *Mastacomys fuscus* HB 140-195mm T 95-135mm

G. Hoye

The only member of its genus, this rodent has a broad, short head, short ears, a compact body, thick fur and a relatively short, barely furred tail, narrow at the base. It inhabits sub-alpine button grass swamps and may spend some of the winter under snow. It usually sleeps by day in dense vegetation, feeding at night on grasses, sedges and seeds. Breeding occurs in summer. The female has four teats and usually rears two young, which bite onto its teats; they become independent at about six weeks.

Grassland Melomys *Melomys burtoni* HB 125-140mm T 125-140mm

R. & A. Williams

Melomys has 15 to 20 species, mostly in Melanesia. The slightly prehensile tail has mosaic-like scalation; the hindfeet are broad and adapted to climbing. The Grassland Melomys, largely terrestrial, has long, soft fur, reddish to brown above and whitish below. The head is broad and short with protuberant eyes. It has short, rounded ears and long vibrissae. By day it sleeps in a globular nest of woven grass, supported between stout grass stems. At night it feeds on stems and seeds of grasses and some berries. Breeding is mostly from late autumn to winter. The female has four teats and usually rears two or three young, which bite onto the nipples until independent at about four weeks.

Species of *Melomys* not dealt with: *M. capensis, M. rubicola.*

Fawn-footed Melomys *Melomys cervinipes* HB 95-200mm T 115-210mm

H. & J. Beste

Very similar to the Grassland and Cape York Melomyses, this is reddish-brown to orange-brown above, whitish to grey below. The tail is a little longer than the body. It inhabits well-watered coastal forests, with dense ground cover. By day it sleeps in spherical nest of leaves or grass, in a tree or attached to grass stems. At night it feeds mostly in trees on leaves, shoots and fruits. Most matings occur in spring and summer. The female has four teats and usually rears two young which bite onto a teat and are dragged behind their mother until seven to 10 days old, becoming independent at about three weeks.

Black-footed Tree-rat *Mesembriomys gouldi* HB 250-320mm T 310-415mm

H. & J. Beste

 The two species of this genus are large, arboreal, and have a rat-like head with large (but not protuberant) eyes, long, rounded ears, broad hindfeet with long claws and a slender tail, much longer than the head and body and with a brushed tip. The Black-footed Tree-rat has black or mottled feet and a white brush to the tip of the tail. It inhabits tropical dry sclerophyll forest and woodland, sleeping by day in a tree-hole. At night it feeds on fruits, flowers and large seeds. Most births are in winter. The female has four teats and usually rears two young, which bite onto its teats until about three weeks old, becoming independent at about four weeks.

Species of *Mesembriomys* not dealt with: *M. macrurus.*

House Mouse *Mus musculus* HB 60-95mm T 75-95mm

Dick Whitford

With some 40 species in Eurasia and Africa, the genus *Mus* is so variable that it can only be defined by a cluster of cranial, dental and biochemical characters, otherwise, it must be said to be 'mouse-like'. The House Mouse, now the most common and widespread rodent in Australia, is the 'typical' mouse, compact-bodied, with a short head, large, rounded ears, small eyes, and a long, slender, scaly tail. It can be distinguished from all native Australian rodents by a notch on the inner surface of the upper incisors. It lives in almost any environment from rainforest to desert, and in buildings, spending the day in a roughly constructed nest of any available soft materials. In rural areas its density is limited by the availability of food and by sufficiently moist soil to make a nesting burrow. At night it feeds on seeds, fruit, food scraps and, where available, cereals; in some circumstances it eats insects. When several years of adequate rainfall leads to soft soil and abundant cereal crops, populations of the House Mouse may irrupt into plagues. After devastating bushfires the House Mouse is usually the first mammal to colonise areas of regrowth. Breeding can occur at any time of the year and is largely determined by rainfall. The female, which becomes sexually mature at the age of about eight weeks, has 10 teats and rears four to eight young in a litter. These become independent at about 18 days.

Spinifex Hopping-mouse *Notomys alexis* HB 95-115mm T 130-150mm

H. & J. Beste

Mice that hop have evolved independently on other continents, the jerboa of North Africa being the best known. The genus *Notomys*, restricted to Australia, is characterized by long, slender hindlegs and a long, slender tail with a brushed tip. The eyes are large and protuberant, the ears large and sparsely furred, the vibrissae on the snout are long. Nine species are recognized, four of which became extinct in the nineteenth century; one is very rare. The Spinifex Hopping-mouse, which has a wide distribution in sandy parts of the central deserts, is brownish above, greyish-white below, has relatively short ears and very long vibrissae. Both sexes have a glandular pouch on the throat. A social species, it sleeps during the day in a deep, elaborate burrow system with numerous inconspicuous pop-holes, usually in the company of other individuals and their dependent young. At night it feeds on seeds, roots, shoots and insects. It does not need to drink. Breeding can occur at any time of the year, but this seems to be mostly in response to rainfall. The female has four teats and usually rears four young, which become independent at the age of about four weeks.

Species *Notomys* not dealt with: *N. aquilo, N. fuscus.*

Fawn Hopping-mouse *Notomys cervinus* HB 90-120mm T 100-160mm

Dick Whitford

This hopping-mouse inhabits gibber plains. It has a pinkish tinge to its upperparts and is white below. The eyes are large and protuberant, ears long and rounded, vibrissae dark and long. The male has a glandular patch on the throat. The Fawn Hopping-mouse inhabits low shrubland and tussock grassland, making a shallow, communal burrow in stony and clay soils. At night it feeds on seeds and green vegetation. It does not need to drink. Breeding is in response to rainfall. The female has four teats and usually rears three young, which become independent at four weeks.

Mitchell's Hopping-mouse *Notomys mitchelli* HB 100-125mm T 140-155mm

H. & J. Beste

This is the largest of the hopping-mice, fawn to dark grey above and greyish-white below, with long, oval ears and a dark brush to the tip of the tail. It lacks glandular pouches on the throat. Inhabiting sandy shrubland, it spends the day communally in a deep burrow system with several pop-holes. At night it feeds on seeds, leaves, roots and insects. It needs to drink. Breeding is in response to rainfall. The female has four teats and rears two to four young, which become independent at about five weeks.

Prehensile-tailed Rat *Pogonomys mollipillosus* HB 130-150mm T 160-210mm

H. & J. Beste

The primarily New Guinean genus includes three or four species, characterized by dense, woolly fur, short, rounded ears, and a long, slender, scantily-haired, scaly, prehensile tail that curls upwards (rather than downwards, as in most mammals). The Australian population (the identity of which is not certain) spends the day in a communal burrow, feeding at night on the ground and in trees, on leaves and nuts. The female has six teats and usually rears two or three young in a litter.

Delicate Mouse *Pseudomys delicatulus* HB 50-80 mm, T 50-80mm

G. Hoye

With at least 20 species, *Pseudomys* is the largest genus of Australian rodents. Most are 'mouse-like' but larger species are referred to as 'rats'. All have a long, smoothly-haired tail. The Delicate Mouse, half the weight of the House Mouse, has smooth fur and large, rounded ears and protuberant eyes. Widely distributed, it sleeps by day in a burrow, feeding at night on seeds. Breeding occurs in mid-winter. The female has four teats and rears two to four young.

Species of *Pseudomys* not dealt with: *P. apodemoides, P. bolami, P. desertor, P. fieldi, P. gouldii, P. higginsi, P. johnsoni, P. laborifex, P. nanus, P. novaehollandiae, P. occidentalis, P. oralis, P. pilligaensis.*

Ash-grey Mouse *Pseudomys albocinereus* HB 60-100mm T 80-110mm

Dick Whitford

A little larger than the House Mouse, this has soft fur, silver-grey above (tinged with fawn), white underparts, pink paws, large ears amd a pale tail. It inhabits semi-arid heathland, sleeping by day in a deep burrow system which accommodates at least a breeding pair and their dependent young. At night it feeds on green plants, seeds and insects. Breeding is mostly in spring. The female, which is smaller, has four teats and usually rears four young, which become independent at about four weeks.

Plains Rat *Pseudomys australis* HB 100-140mm T 80-120mm

K. Atkinson

About 50% heavier than the House Mouse, the Plains Rat has thick, soft fur, greyish-brown above and whitish below, relatively large ears and a tail that is much shorter than the head and body and pale at the tip. It inhabits tussock or hummock grassland, often on gibber plains and spends the day communally in a complex burrow system. At night it feeds mainly on seeds. It does not need to drink. Breeding is in response to rainfall. The female has four teats and usually rears four young, which become independent at about four weeks. The female is mature at about 10 months and may rear up to three litters in a year.

Western Pebble-mound Mouse *Pseudomys chapmani* HB 50-70mm T 70-80mm

M. Gillam

A little smaller than the House Mouse, this species is distinguished by a rather long, 'roman- nosed' snout, buff-brown upperparts, a dark head, white underparts and lower lip. It inhabits pebbly soils in arid tussock grassland and acacia woodland, spending the day in a nest made within a large structure of pebbles, brought to a site in the mouths of successive generations. As well as providing shelter, pebble-mounds may act as dew-traps, concentrating moisture in the nesting area.

Smoky Mouse *Pseudomys fumeus* HB 85-100mm T 100-145mm

R. W. Jenkins

Four to five times the weight of the House Mouse, this species has soft, dark grey fur on the upper part, grey-white below, a dark muzzle, a dark ring around the large eyes and pink feet. The tail, distinctly grey above and white below, is much longer than the head and body. It inhabits heath understorey of dry sclerophyll forest and woodland in sub-alpine areas. By day it sleeps in a shallow burrow. In summer it feeds on seeds and berries; in winter more on underground fungi; in spring, on Bogong Moths. Breeding occurs in summer. The female has four teats and rears three or four young in a litter.

95

Eastern Chestnut Mouse *Pseudomys gracilicaudatus* HB 100-115mm · T 80-120mm

A. Young

Up to four times the weight of the House Mouse, this bulky, short-faced mouse has thick, rather bristly, grizzled chestnut hair on the upperparts, greyish below. The face is very short, the eyes protuberant with a pale rim, the ears short and rounded; hairs on the hindfeet extend beyond the claws. It inhabits heathland and woodland with dense understorey and spends the day in a nest, constructed above ground or in a shallow burrow. It feeds at night on grasses and seeds. Breeding is from August to March. The female has four teats and usually rears three young, which become independent at about four weeks.

Sandy Inland Mouse *Pseudomys hermannsburgensis* HB 60-90mm T 70-90mm

Babs & Bert Wells

Resembling the House Mouse, this species is a little smaller, has larger ears and eyes, and lacks the typical musty odour of that species. The head is blunter. It inhabits hummock grassland and mulga scrub. By day it sleeps communally in a nest in a deep burrow, emerging at night to feed on seeds, shoots, roots and small tubers. Breeding is in response to rainfall. The female has four teats and usually rears three to four young, which become independent at about 30 days.

Heath Mouse *Pseudomys shortridgei* HB 90-120mm T 80-110mm

H. & J. Beste

About four times the weight of the House Mouse, this has long, dense fur, flecked grey-brown above, paler below. The ears are short and rounded, partially covered by the hair of the head. The face is blunt and the eyes large. The tail, which is shorter than the head and body, is well-furred. It is sometimes referred to as the "Heath Rat". It inhabits cool heathland and is most prevalent in the early stages of vegetation regrowth after bushfires. As vegetation returns towards a climax, its numbers decrease and it becomes locally rare or even disappears. It usually rests during the day in a short burrow (sometimes made by other species) but may build a nest on the ground. At night it forages terrestrially for a variety of foods. In the warmer part of the year it eats flowers, fruits and berries; in autumn it takes the leaves and stems of grasses, sedges and lilies; in winter it depends largely upon subterranean fungi. Breeding occurs from November to February. The female has four teats and usually rears two to three young. Longevity of about four years, production of two or three litters a year, and aggressive dispersal of young provide a constant pressure to explore newly created habitats.

Bush Rat *Rattus fuscipes* HB 110-195mm T 105-200mm

G. Hoye

More than 50 species are recognized in the genus *Rattus*, distributed through Europe, Asia, Indonesia and Melanesia. Seven species are native to Australia, two have been introduced by Europeans. The genus is difficult to define, even on cranial and dental criteria, but the Australian species share the following characters: relatively large eyes, thin, rounded ears, a 'roman nose' accentuated by long hair between the eyes, a sparsely haired, slender tail with rings of scales, about the same length as the head and body; long, usually rather coarse, body hair. The Bush Rat has relatively soft fur, prominent ears, pale forefeet, and a tail that is a little shorter than the head and body. It inhabits tropical to cool temperate rainforest and wet sclerophyll forest with a dense understorey, sleeping by day in a rough nest in a burrow under a rock or fallen timber. At night it feeds on a wide variety of green plants and fungi, also taking a large number of terrestrial insects and their larvae. It is solitary. Breeding occurs throughout the year, with a peak of births in summer. The female, which is smaller than the male, has eight to 10 teats and usually rears about five young, which become independent when about four weeks old. The female becomes sexually mature when about 12 weeks old.

Species of *Rattus* not dealt with: *R. colletti*

Cape York Rat *Rattus leucopus* HB 130-210mm T 135-210mm

H. & J. Beste

Also occurring in New Guinea, this rainforest species has fawn fur and a rufous or golden tinge to the upper parts, the muzzle is much more pointed than in other Australian rats, the hindfeet are white, the tail often mottled with pale patches and slightly longer than the head and body. Its short burrow has several interconnected chambers. At night it feeds on terrestrial insects, fungi, fruits and nuts. Breeding has a peak in summer. The female has six (sometimes eight) teats and usually rears between two to five young, which become independent at about 25 days.

Swamp Rat *Rattus lutreolus* HB 130-200mm T 80-150mm

G. Hoye

This has rather coarse fur, grey-brown above, the eyes are not protuberant, the dark, sparsely furred tail is about two-thirds the length of the head and body, the short hindfeet are dark grey. It inhabits grassland, sedgeland and heath in well-watered areas. By day it sleeps in a short burrow or cavity. Where the substrate is wet it makes an elevated nest of woven vegetation. At night it feeds on grasses, sedges and terrestrial arthropods. Breeding is in spring and summer. The female has 10 teats (eight in Tasmania) and usually rears three or four young, which become independent at three to four weeks.

Brown Rat *Rattus norvegicus* HB 180-255mm T 150-215mm

C. Henley

This introduced rodent is grey-brown above, with somewhat bristly fur; the snout tapers but ends bluntly; the ears are relatively large but do not reach the eyes when folded forward; the naked tail is rather stout and shorter than the head and body. It lives in social groups, spending the day in a nest usually associated with human habitations. It climbs well but feeds mostly at ground level. Breeding is continuous. The female has 12 teats and usually rears seven to eight young, which become independent at about three weeks.

Black Rat *Rattus rattus* HB 160-210mm T 180-250mm

A. Young

Significantly smaller than the Brown Rat, this is darker grey above, the ears are large, extend beyond the head fur and reach past the middle of the eye when folded forward, the snout tapers to more of a point than that of the Brown Rat. The naked tail is longer than the head and body and has overlapping scales. It inhabits well-watered areas, usually in association with human habitations. By day it sleeps in a communal nest. At night it feeds on food scraps, stored food, seeds, nuts, fruits and insects. Breeding is continuous. The female has 10 or 12 teats and usually rears five to 10 young, which become independent at about three weeks.

Canefield Rat *Rattus sordidus* HB 110-210mm T 100-200mm

G. Hoye

This native rat has rather coarse hair, blackish-brown above; the large ears are not covered by the head fur, the head has a 'roman-nosed' appearance, soles of the hindfeet are pale. Preferring tropical grasslands, it thrives in cane-fields and feeds at night. By day it sleeps communally in a burrow. Breeding has a peak in autumn. Under favourable conditions populations irrupt into plagues. The female has 12 teats and usually rears about six young, which become independent at about three weeks.

Pale Field-rat *Rattus tunneyi* HB 120-200mm T 80-150mm

Babs & Bert Wells

This rat is distinguished by long, brindled yellowish-brown fur, paler than in other Australian rodents, large, protuberant eyes, relatively large ears; somewhat tapering muzzle and pale feet. It inhabits well-watered tropical and subtropical tall grassland. By day it sleeps in a nest in a shallow burrow system with several pop-holes. At night it feeds on grass stems, seeds and roots. Breeding occurs in autumn in Queensland, somewhat later in the Northern Territory and Western Australia. The female has 10 teats and usually rears four or five young, which become independent at three weeks.

Long-haired Rat *Rattus villosissimus* HB 120-220mm T 100-180mm

Babs & Bert Wells

This large rodent (sometimes called the 'Plague Rat') resembles the Canefield Rat but has lighter grey fur, with longer, dark guard hairs, and lacks the brown to rufous flank colouration of that species, the tail is rather well-furred with dark hairs and is often shorter than the head and body. The only member of the genus that is adapted to arid conditions, it occurs throughout most of the hotter parts of the Central Desert but varies immensely in population density. It sleeps during the day in a short burrow, usually dug under shelter and with several pop-holes. At night it forages for grasses, herbs and seeds. Usually very sparsely distributed, it breeds close to its full reproductive potential after several years of good rain, irrupting into a very noticeable plague in response to which there is a dramatic upsurge in numbers of owls and hawks. The female, which is about two-thirds the weight of the male, has 12 teats and usually rears six or seven young, which become independent at about three weeks. After the explosive expansion of a population, the species reverts to a sparse and very scattered distribution, presumably surviving in various refuges.

White-tailed Rat *Uromys caudimaculatus* HB 250-365mm T 240-360mm

Dick Whitford

Sometimes referred to as the Giant White-tailed Rat, this native rodent is one of two members of its genus in Australia; several others occur in New Guinea. About the size of a Rabbit, it has coarse hair, grey-brown above, cream below, and pale paws. The long, naked tail is black at the base, otherwise white, and is about the length of the head and body; it is slightly prehensile. It is arboreal and spends the day in a nest in a tree-hole, emerging at night to forage opportunistically for nuts, fruits, insects and small vertebrates. Although its normal habitat is wet tropical forest, it often enters camps and houses in search of food. Breeding occurs throughout the year. The female, which is smaller than the male, has four teats and usually rears two or three young, which attach themselves firmly to its teats and are dragged with it as it moves about. They are weaned when about five weeks old but remain dependent until the age of about eight weeks, when the female becomes sexually mature.

Species of *Uromys* not dealt with: *U. hadrourus.*

103

Common Rock-rat *Zyzomys argurus* HB 85-120mm T 100-130mm

G. Hoye

The five species of this endemic genus have a rather blunt 'roman nose;' relatively large ears; large protuberant eyes, somewhat spiny fur, and a naked, rather thick-based tail. The Common Rock-rat, which inhabits open forest to grassland has coarse, golden hair on the upper parts, short, rounded ears, large, protuberant eyes and a broad-based tail slightly longer than the head and body and with a small terminal brush. By day it sleeps in a

nest in a crevice in a rock-pile, emerging at night to feed on leaves, seeds, plant stems, fungi and insects. Breeding peaks from autumn to early winter. The female has four teats and rears up to four young, which become independent at about four weeks.

Species of *Zyzomys* not dealt with: *Z. pedunculatus, Z. palatalis, Z. maini*.

Kimberley Rock-rat *Zyzomys woodwardi* HB 104-169mm T 94-135mm

H. & J. Beste

Larger than the Common Rock-rat, this is cinnamon brown above, white below, with a moderately furred tail, thick at the base (often foreshortened). It is smaller than other species, except the Common Rock-rat, from which it is distinguished by cranial characters. It is restricted to northern Australian monsoon forest, sclerophyll forest and scrubland, over rocky slopes, scree and rock-piles. By day it sleeps in a rock crevice, emerging at night to feed on the fallen seeds of rainforest trees and perennial grasses. Breeding is continuous but least in the dry season. The female has four teats, to which the young cling for at least two weeks, becoming independent at about four weeks.

Bare-backed Fruit-Bat *Dobsonia moluccensis* HB 280-320mm T 20-30mm

Queensland Museum

There are up to 15 species of this genus from the Solomon Islands to the Philippines, characterized by the attachment of the naked wing-membranes along the mid-line of the back, covering the fur and presenting a bare-backed appearance. There is no claw on the second digit of the forelimb. What is known in Australia and New Guinea as the Bare-backed Fruit-Bat is a typical member of the genus and differs from all other Australian bats in the characters mentioned above. It is brownish-black and has an elongated snout with prominent nostrils. It inhabits tropical rainforest and roosts during the day in a tree or in the twilight zone of a cave. In Australia it feeds on a wide variety of rainforest fruits and flowers and on the nectar and blossoms of eucalypts. It is able to fly fast, but also to hover and even fly slowly backwards. Mating occurs from March to May. A single young is born in October or November and appears to be nursed for five to six months. The female becomes sexually mature when about two years old.

Northern Blossom-bat *Macroglossus minimus* HB 60-70mm T minute

G.C. Richards

This genus comprises three species. The Northern Blossom-bat, which is the only species represented in Australia, has russet fur on the upper parts, a long snout, long tongue, broad, rounded ears and a rudimentary web between the tail and hindlegs. It differs from the Queensland Blossom-bat, by a large gap between its small lower incisors. It inhabits monsoon forest and tropical woodland, and rests by day singly or in small groups, in dense foliage. At night it flies slowly in the canopy feeding on nectar and pollen, lapped up with its long tongue. Breeding is continuous. A single young is born.

Eastern Tube-nosed Bat *Nyctimene robinsoni* HB 100-110mm T 20-30mm

G.B. Baker

Up to 14 species of this genus occur from Sulawesi to the Philippines, characterized by protuberant, tubular nostrils, a short, rounded snout, a short tail, and absence of claws from all but the first digit of the forelimb. The Queensland Tube-nosed Bat has yellow spots over the wings and ears. It inhabits rainforest to dry sclerophyll forest and woodland. By day it roosts singly in dense foliage. At night it feeds on flowers and fruits. The female gives birth to a single young between October and December and carries it until quite large.

Eastern Blossom-bat *Syconycteris australis* HB 40-60mm T minute

The genus *Syconycteris* extends from the Moluccas to the Bismarck Archipelago. It comprises two species, one of which occurs in New Guinea and coastal Queensland. The Eastern Blossom-bat (also known as the Common Blossom-bat) is very similar to the Northern Blossom-bat but distinguished from it by not having a large gap between the two lower incisors. Mouse-sized, it weighs only about 15g, and is specialised for a diet of nectar and pollen. Nectar is lapped directly with a long, protrusible tongue, pollen is collected on the fur and is ingested by grooming with the teeth. It flies slowly

and is able to hover, like a hummingbird, but not sufficiently to permit feeding; instead, it alights forcibly on or near a blossom. It is solitary and roosts during the day in dense canopy foliage, emerging at night to forage over a radius of a few kilometres. In tropical parts of the range, breeding is continuous. In New South Wales a single young is born in October or November, another in February or April. It is suckled for about three months before becoming independent.

Grey-headed Flying-fox *Pteropus poliocephalus* HB 230-290mm T minute

G.B. Baker

This very successful genus includes some 60 species from Zanzibar to the Cook Islands. Four species are known from Australia, another two await description. It is characterized by large size, a fox-like snout, large eyes, simple nostrils and ears, claws on the first two digits of the forelimb; the tail is vestigial. The Grey-headed Flying-fox has a white or greyish head, a reddish-yellow mantle around the neck, and thick, shaggy fur which extends to the ankles. It is common in rainforest and well-watered sclerophyll forest over much of the eastern and southeastern mainland, usually in large, partly nomadic colonies. It roosts communally by day in 'camps' that are usually situated close to water in gullies with a relatively dense forest canopy. At night it forages as far as 50km in search of fruits of rainforest trees and the blossoms of eucalypts, banksias and tea-trees. Mating can occur at any time of year but most births occur about October. The single young is carried by the mother until about four to five weeks old, thereafter left in the camp while the mother forages.

Black Flying-fox *Pteropus alecto* HB 240-260mm T minute

Largest and one of the most familiar of the Australian flying-foxes, this species is characterized by a covering of short, black hair. It is sometimes reddish around the neck and above the brown eye-rings. It is widely distributed in coastal tropical Australia (also from Sulawesi to New Guinea), usually in mangrove or melaleuca swamps, sometimes in patches of wet sclerophyll or rainforest. It roosts by day in dense canopy, usually in camps of some thousands, often in the company of other species. At night it travels as far as 50km from the camp in search of the flowers and fruits of eucalypts and rainforest trees. It can cruise at a speed of 25 to 35 km/h. In the northern part of its range mating occurs throughout the year; in March or April in the more southern parts. In the south a single young is born between September and November and is carried by the mother clinging to the teat in its armpit until about one month old. It is suckled in a creche until reaching independence at the age of about three months.

Spectacled Flying-fox *Pteropus conspicillatus* HB 220-240mm T minute

G.B. Baker

The body fur of this flying-fox is blackish, sometimes frosted; pale eye-rings give the impression of 'spectacles'. Also occurring in the New Guinea region, it inhabits tropical rainforest, wet sclerophyll forest and mangroves. By day it roosts in camps of up to hundreds of thousands, usually within rainforest. At night it ranges widely in search of pale-coloured fruits of rainforest trees. It does not feed on nectar and pollen. It drinks by skimming the surface of a body of water and sometimes takes sea water in this manner. Mating is from March to May and a single young is born about October or November.

Little Red Flying-fox *Pteropus scapulatus* HB 190-240mm T minute

G.B. Baker

Smallest of the Australian flying-foxes, this has short reddish-brown fur over most of the body and is brown to yellow on the shoulders and around the eyes. The hindlimbs are bare and only sparsely furred. It is widespread in regions ranging from rainforest to sub-arid woodland, roosting by day in tall trees in camps of up to a million individuals. At night it feeds on nectar and pollen of eucalypts, also on fruits, leaves, shoots, sap and insects. Mating is from November to January; most births take place in April or May. The single young is carried for a month and suckled in a creche until about four months old.

Ghost Bat *Macroderma gigas* HB 100-130mm T minute

G. Anderson

Macroderma is one of four similar genera in the family Megadermatidae, extending from Africa to Asia, characterized by large size, long, erect ears, partially joined above the forehead, relatively large eyes; and an elaborate nose-leaf. These characters separate the Ghost Bat from all other Australian bats. In contrast with other microbats, it employs both vision and echolocation in its navigation and predation. It is widely, but patchily, distributed in tropical and subtropical Australia in environments ranging from rainforest to arid woodland, roosting communally during the day in a cave or under a rock-shelf. At night it is a predator on other bats, taking them in the air; and terrestrial vertebrates, swooping upon these, enfolding them in its wings and killing them with powerful bites. Prey is taken to a feeding site to be eaten and these sites are marked by an accumulation of the uneaten remains. Mating appears to occur mainly in July or August. A single young is born between September and November in a maternity colony separate from those of the male. Young are suckled in a creche and later fed with prey brought back by the mother. Juveniles accompany their mothers on hunting expeditions.

Eastern Horseshoe-bat, *Rhinolophus megaphyllus* HB 42-59mm T 38-43mm

Dick Whitford

The genus *Rhinolophus* comprises some 70 species distributed through Europe, Africa, Asia and Melanesia. Members are known as horseshoe-bats in reference to the shape of the complex nose-leaf, similar to that of the leafnosed bats of the genus *Hipposideros*. *Rhinolophus* has the more complex nose-leaf with a vertical component, the 'lancet'. The toes have three joints. The female has a false teat in the groin, to which the young attach while dependent. Two species occur in Australia, both also represented in Melanesia. The Eastern Horseshoe-bat is grey-brown and is very similar to the Greater Horseshoe-bat, but smaller and with shorter ears (less than 20mm) and forearms (less than 50mm). Some individuals are rufous. It is distributed over most of the eastern coast of Australia where it roosts communally in caves, rock-piles and equivalent artificial structures with a humid microclimate, in environments ranging from cool rainforest to woodland. At night it flies slowly above the forest understorey, feeding on relatively large insects. It is able to hover while gleaning on insects from foliage. Mating is usually from April to June. The females form a maternity colony where each gives birth to a single young around November. It is carried by the mother for a month, thereafter suckled in the roost until becoming independent at the age of about six weeks.

Greater Horseshoe-bat *Rhinolophus philippinensis* HB 62-65mm T 25-35mm

This is very similar to the Eastern Horseshoe-bat but larger, with longer ears (more than 20mm) and forearms (more than 50mm). It occurs from Sulawesi to the Philippines and, in Australia, is restricted to coastal northern Queensland (the Eastern Horseshoe-bat extends to eastern Victoria). It is not very social, often roosting singly in a cave or mine, sometimes in the company of the more numerous Eastern Horseshoe-bat. At night it flies slowly, feeding on relatively large insects taken in the understorey of a forest or over water. It can hover, then dart at its prey. Young are probably born in late October or November.

G.B. Baker

Dusky Leafnosed-bat *Hipposideros ater* HB 40-50mm T 20-30mm

There are some 50 species *Hipposideros* from Africa to the Philippines: five occur in Australia. Sometimes known as horseshoe-bats, they are here called leafnosed-bats, distinguished from *Rhinolophus* by a somewhat squarer nose-leaf, absence of a

'lancet' in the nose-leaf, and presence of two joints in the toes. The Dusky Leafnosed-bat is greyish (sometimes ginger) above, the ears are triangular and bluntly pointed, the nose-leaf lacks accessory leaflets. Some individuals are reddish-golden. It roosts, usually communally, in a cave or tree-hole. At night it flies slowly in or above the forest understorey, searching for large insects. One young is born in a maternity colony in October or November, where it is suckled until independent at about six weeks.

G.B. Baker

Species of *Hipposideros* not dealt with: *H. cervinus*, *H. stenotis*.

113

Diadem Leafnosed-bat *Hipposideros diadema* HB 75-85mm T 30-40mm

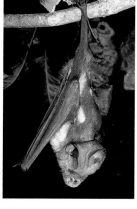

G. Hoye

Much larger than any of the other Australian leafnosed bats, this has grey to yellowish-brown fur, often with pale patches. There are three accessory leaflets under each side of the lower nose-leaf; some individuals are bright orange. Extending from the Nicobar Islands to the Philippines, it occurs in the Northern Territory and in north-eastern Queensland. It roosts, usually singly, in a cave or buildings. At night it flies more directly and with less hovering than in other leaf-nosed bats, in search of large flying insects. The females form maternity colonies and the single young is born around November or December.

Northern Leafnosed-bat *Hipposideros semoni* HB 40-50mm T 10-20mm

G.B. Baker

This little-known bat, which also occurs in New Guinea, is solitary. It is brown above, paler below, and has a club-shaped protuberance projecting from the centre of the lower portion of the nose-leaf and a smaller structure ('wart') on the centre of the upper portion. It roosts singly in a cave, rock crevice, or a variety of confined spaces in artificial structures. At night it flies slowly and with a fluttering wing-beat, above and within the understorey of tropical rainforest or woodland, searching for large flying insects. Little is known of its reproduction and development.

Orange Leafnosed-bat *Rhinonicteris aurantius* HB 40-60mm T 20-30mm

G.B. Baker

This endemic genus, closely related to the leaf-nosed-bats, has a rich, golden, silky fur on the upperparts; small, sharply pointed ears and a complex nose-leaf, the lower part of which is horseshoe-shaped, the upper part scalloped. The single species inhabits tropical and subtropical sclerophyll forest and woodland, roosting communally (groups of tens to thousands) in warm, humid caves, often in the company of the Dusky Leafnosed-bat. At night it forages below the canopy in search of beetles, bugs, ants, weevils and other insects, some gleaned from foliage. Prey is often taken back to the roost to be eaten.

Coastal Sheathtail-bat *Taphozous australis* HB 80 -90mm T 20-25mm

G.B. Baker

Taphozous contains 13 species from Africa to the Philippines, with four species in Australia. Most have a glandular pouch on the throat. The Coastal Sheathtail-bat resembles the Yellow-bellied Sheathtail-bat but has narrower ears; the belly is much the same colour as the upperparts. The tail appears to pierce the membrane between the hindlegs. The Coastal Sheathtail-bat is smaller than the Common Sheathtail-bat and is either grey or brown. The throat-pouch is rudimentary in the female. It roosts in colonies in caves, seldom more than 2km from the coast. At night it hunts slowly above the canopy of dune scrubland and melaleuca swamps. A single young is born around October or November.

Species of *Taphozous* not dealt with: *T. hilli, T. kapalgensis.*

Common Sheathtail-bat *Taphozous georgianus* HB 60-90mm T 20-30mm

G.B. Baker

This Sheathtail-bat is dark brown above, pale brown below, with yellow-brown hairs under the base of the tail. Neither sex has a throat-pouch. It is widespread in mainly arid to semi-arid country from Queensland to the Western Australian coast. It roosts, usually in groups of two to 20, near the entrance to a cave, or abandoned mine, clinging to the wall ('spider-like') with four limbs. If disturbed it scuttles over the wall or roof of the cave, seeking shelter in a rock crevice. It does not hibernate, but when food is scarce and it has utilised most of the fat reserves laid down in autumn, it may allow its temperature to fall and can become quite slow in its reactions. At night it flies fast over vegetation and bodies of water, taking beetles and other flying insects, that are eaten in flight.

Individuals appear to have particular foraging areas that are systematically patrolled each night. Mating usually occurs in August or September. The female gives birth to a single young in about December. It attaches itself to a teat and is carried by the mother for three to four weeks (weighing up to half the weight of the mother) after which it is able to fly. Adult size is reached at the age of three months: the females becomes sexually mature when nine months old.

Yellow-bellied Sheathtail-bat *Saccolaimus flaviventris* HB 70-90mm,
T 20-35mm

Babs & Bert Wells

Sheathtail-bats, which also include the genus
Taphozous, have a tail that appears to penetrate
the membrane between it and the legs, protruding through above it. Of the five species of *Saccolaimus*, three occur in Australia. The Yellow-
bellied Sheathtail-bat is distinguished by glossy
black fur on the upperparts, contrasting with white or cream underparts. When at rest, the tips of the long wings are folded back over
the rest of the membrane. The muzzle is flat and sharply pointed.
The male has a prominent throat-pouch with a glandular region
behind it. It roosts in tree-holes in a very wide range of environments from rainforest to arid shrubland, usually singly, but sometimes in groups of up to 10. It is sometimes found on the walls of
buildings. It flies fast and with agility, above the canopy or in
clearings, in pursuit of relatively large flying insects such as beetles and moths. The female produces two young between September and March. Populations in the southern part of the range probably migrate northward in winter.

Species of *Saccolaimus* not dealt with: *S. mixtus, S. saccolaimus.*

Little Freetail-bat *Mormopterus loriae* HB 47-55mm T 30-36mm

G.B. Baker

The genus *Mormopterus*, sometimes known as 'mastiff-bats', includes 10 species in South America, Africa, Melanesia and Australia, where four species occur. Like other members of the family, Molossidae, it is characterized by a wrinkled 'mastiff-like' muzzle and a tail that projects beyond the membrane between the hindlegs. The four Australian species differ from other Australian members of the family in having a less wrinkled muzzle, a less hairy face, and pointed ears that are not joined at their bases. Most species lack a throat-pouch. They are difficult to distinguish

from each other but have distributions that overlap only slightly. The Little Free-tail Bat, which also occurs in New Guinea, occupies a variety of environments from tropical sclerophyll forest to woodland. It roosts communally in tree-holes, crevices, and the roofs of unoccupied buildings. Its wings are short, narrow and pointed. At night it flies fast and with agility, above and alongside the forest canopy, over scrubland and over water, in search of flying insects. It may alight on plants or on the ground to take flightless insects. The female gives birth to a single young from November to January.

Species of *Mormopterus* not dealt with: *M. beccarii, M. planiceps*

Eastern Freetail-bat *Mormopterus norfolkensis* HB 40-60mm, T 35-45mm

G.B. Baker

One of the smaller Australian freetail-bats, this has long fur and more wrinkled tail and wing-membranes. The male has a throat-pouch. It inhabits warm temperate to subtropical rainforest and sclerophyll forests and woodland (but does not occur in Norfolk Island). Little is known of its biology but it appears to be a fast-flying predator, below the canopy and over water. It roosts, singly or in small colonies in tree-holes, rock crevices, or under roofs. At night it flies directly above and below the canopy in search of flying insects. Nothing is known of its reproduction or development.

White-striped Mastiff-bat *Nyctinomus australis* HB 80-110mm T 40-60mm

G. Hoye

The genus *Nyctinomus*, with one species restricted to Australia, is distinguished from other Australian molossids by a very wrinkled muzzle, close approximation (without fusion) of the bases of the ears, and well-developed throat-pouches in both sexes. There is a white stripe between the upper and lower surfaces. The White-striped Mastiff-bat roosts singly or in small groups in tree-holes or similar crevices. At night it forages above the canopy, particularly along the tree-edges of waterways, in search of flying insects. Large insects are taken on the ground. The female produces a single young around December.

Gould's Wattled-bat *Chalinolobus gouldii* HB 60-80mm T 40-50mm

Dick Whitford

Chalinolobus includes seven species, from New Guinea to New Zealand, with five in Australia. It has lobes ('wattles') at the sides of the lower lip, the outer margin of the ear is close to the margin of the mouth. Gould's Wattled-bat is black above and black or brown below and distinguished by having the outer margin of the short, broad ear extended into a prominent lobe. It ranges from tropical rainforest to arid woodland, particularly along watercourses. It roosts in tree-holes or similar spaces. At night it forages below the canopy for slow-flying insects. Caterpillars are gleaned from vegetation. In the south this bat becomes torpid in winter. Mating is about May. Twins are born near or during November and carried by the mother for some time, becoming independent at about six weeks.

Species of *Chalinolobus* not dealt with: *C. picatus.*

Chocolate Wattled Bat *Challinolobus morio* HB 50-60mm T 40-50mm

A. Young

This is distinguished from other Australian wattled-bats by chocolate brown fur on the upper and lower parts. Its wattles are small and the forehead is domed. It inhabits a range of environments from wet sclerophyll forest to semi-arid scrubland, usually roosting in groups of 20 to several hundred individuals in tree-holes, caves, and a wide variety of artificial structures. In the southern part of its range, it is the last of the bats to enter hibernation. One young is born in November or December in the northern part of the range, earlier in the south.

Hoary Wattled-bat *Chalinolobus nigrogriseus* HB 40-60mm T 30-40mm

Also occurring in New Guinea, this wattled-bat has white tips to its woolly body fur, it has shorter ears than the Large-eared Pied Bat. It roosts singly or in small groups in rock crevices or tree hollows. At night it flies with great manoeuvrability, below the forest canopy or over shrubs, feeding on moths and other flying insects. It also gleans some non-flying insects from the bark of trees or from the ground. The female gives birth to two young, probably about December.

G. Hoye

Large-eared Pied Bat *Chalinolobus dwyeri* HB 47-53mm T 42-46mm

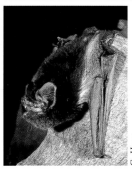

This wattled bat is characterized by glossy black fur with a white fringe around the body and beneath the wings and tail-membrane. It has larger ears than the Little Pied Bat. Relatively recently recognized as a distinct species, it ranges from coastal wet sclerophyll forest to dry sclerophyll and open woodland. It roosts in small groups in the twilight zone of caves or mines. At night it forages below the canopy for small flying insects. It is likely that in winter, in the southern part of the distribution, it hibernates; it certainly becomes torpid. The female usually gives birth to twins in November or December.

G. Hoye

121

Golden-tipped Bat *Kerivoula papuensis* HB 50-60mm T 40-50mm

G. Hoye

Some 20 species of this genus, often known as 'painted' or 'woolly' bats, occur from Africa to Australia. The Golden-tipped Bat, also found in New Guinea and sometimes referred to as *Phoniscus papuensis*, is characterized by dark brown fur tipped with gold, and golden hairs on the forearm and upper surface of the hindleg. It appears to fly slowly in dense vegetation, gleaning arthropods (particularly spiders) from vegetation. In New Guinea it is known to roost in dead foliage and in the roofs of houses. In the northern part of its range it probably breeds in summer; elsewhere in the early spring.

Little Bentwing-bat *Miniopterus australis* HB 43-48mm T 43-48mm

G. Little

About 11 species of *Miniopterus* range from Europe to New Caledonia: two are in Australia. The third digit of the forelimb flexes under the upperpart of the wing ('bentwing') when the bat is at rest. The tail is long and completely enclosed between the legs. The Little Bentwing-bat is distinguished by small size and a very long terminal joint in the third digit of the forelimb. It roosts communally in caves or similar spaces, often with the Common Bentwing-bat. At night it forages for small insects beneath the canopy of rainforest, wet and dry sclerophyll forest and paperbark swamps. A single young is born in December and suckled in the roost.

Common Bentwing-bat *Miniopterus schreibersii* HB 52-58mm T 52-58mm

G.B. Baker

The fur on the upper parts of this bat is chocolate brown, slightly paler below. It resembles the Little Bentwing-bat but is larger. Extending from Europe to Australia it is distributed in well-watered parts of eastern and northern Australia, where it roosts in caves, mines, and equivalent artificial excavations, often in thousands. It hibernates in the southern part of its range. At night it flies fast and with agility, feeding on small flying insects. Maternity colonies are formed in spring and a single young is born in late November or early December. It is left in the roost while the mother is feeding.

Tube-nosed Insectivorous Bat *Murina florium* HB 45-50mm T 35-40mm

About 11 species of *Murina* extend through eastern Asia to New Guinea; one extends into Australia. It has long, brown fur on the upper parts, is slightly paler below and distinguished by tubular nostrils. When resting, it folds its wings around the body, like a fruit-bat, but at a little distance from the body, forming an umbrella that diverts condensed water or rain away from the body. It flies slowly over the understorey, feeding on insects, some of which may be gleaned from foliage. It is very rare in Australia and little is known of it here.

G. Hoye

Large-footed Myotis *Myotis adversus* HB 52-56mm T 36-40mm

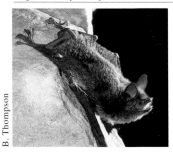

B. Thompson

This genus, with about 100 species, occurs on all continents except Antarctica. It has simple ears, set well apart on the head and unconnected with the corners of the mouth. The Large-footed Myotis, is small and distinguished by its long feet. Its ears are larger than those of bent-wing bats. It roosts communally in caves, similar spaces, and dense rainforest foliage. At night it flies above of rivers and lakes catching aquatic insects and very small fishes with the long claws of its hindfeet. Breeding is continuous in the tropics. In the subtropics, one young is born in October, another in January. In the southern part of the range, one young is born about December.

Greater Broad-nosed Bat *Scoteanax rueppellii* HB 80-95mm T 40-55mm

G. Hoye

This big bat is dark reddish-brown, with a broad and almost bare muzzle, relatively large eyes, and well-separated ears, the outer bases of which come close to the angles of the mouth. It resembles pipistrelles but has two (not four) upper incisors. Inhabiting wettish forests, it roosts in tree-holes and similar spaces. It flies fast and directly, along forest edges or along watercourses, feeding on beetles and other relatively large insects, probably also on smaller bats. Females congregate at maternity sites in suitable trees and each gives birth to a single young in January.

Lesser Long-eared Bat *Nyctophilus geoffroyi* HB 40-50mm T 30-40mm

Dick Whitford

Nine species of this genus extend from New Guinea to Australia, where six occur. The genus is characterized by large ears, usually connected at their bases by a membrane. There is a small horseshoe-shaped nose-leaf on the muzzle. The Lesser Long-eared Bat has light grey-brown upper parts and an elastic membrane between the two lobes at the back of the nose-leaf. It roosts in a wide variety of crevices, in colonies of up to several hundreds. At night it flies slowly, close to the ground, foraging for slow-flying or non-flying insects. Maternity colonies form in spring and twins are born in early summer. These are suckled in the roost.

Species of *Nyctophilus* not dealt with: *N. arnhemensis, N. bifax, N. timoriensis, N. walkeri, N. howensis.*

Gould's Long-eared Bat *Nyctophilus gouldi* HB 55-65mm T 45-55mm

P. German

Similar in appearance to the Lesser Long-eared Bat, this has dark brown to dark grey upperparts and light grey underparts. The nose-leaf is small and the bilobed protrusion on the snout behind the nose-leaf is low. It roosts mainly in tree-holes but also under bark, or in abandoned buildings. At night it flies slowly and with great manoeuvrability, above the understorey of eucalypt forest or woodland, often along watercourses. It takes flying insects but also gleans them from foliage. In the southern part of the range it hibernates. Mating begins in spring. The female gives birth to twins about December.

125

Large Forest Bat *Vespadelus darlingtoni* HB 38-50mm T 31-38mm

L. Lumsden

Vespadelus is regarded by some authorities as indistinguishable from *Eptesicus*. It comprises nine species, all restricted to Australia. The Large Forest Bat is the biggest of these and has a long forearm (32mm or more), and longer, denser fur. It ranges from rainforest to alpine heaths, roosting in tree-holes, and similar crevices, in colonies of up to 50. It hibernates in the southern part of the range. At night it flies fast below the canopy or over heath in pursuit of small insects. A single young born in midsummer is carried by the mother for several weeks then left in the roost while it forages.

Species of *Vespadelus* not dealt with: *V. baverstocki, V. caurinus, V. douglasorum, V. finlaysoni, V. pumilus, V. troughtoni, V. vulturnus.*

Southern Forest Bat *Vespadelus regulus* HB 36-46mm T 28-39mm

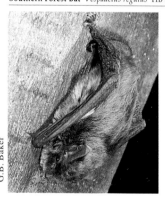

G.B. Baker

Difficult to distinguish from other members of the genus, the Southern Forest Bat is to some extent identifiable by a contrast between the dark brown upperparts and lighter underparts, and a flatter skull. It ranges from wet sclerophyll forest to woodland, roosting in colonies of up to 100 in tree-holes or similar crevices, sometimes with other species. The male roosts separately, except in the mating season. It flies fast and with agility in pursuit of flying insects, particularly moths. It does not take insects from foliage. Mating occurs in autumn and a single young is born in early summer.

Dingo *Canis familiaris* HB 860-980mm T 260-380mm

D.R. & L.K. Corbett

A subspecies (C. f. dingo) of the domestic Dog, the Dingo cannot be reliably distinguished on any external characters. It is often ginger-coloured with white points to the ears and tail, but can be black. It differs from the domestic Dog in that the Dingo breeds only once a year and it seldom barks. Most closely related to the semi-domestic Dog of South-east Asia, it seems to have arrived in Australia about 3 500 years ago. Some Dingoes have a semi-domestic relationship with Aborigines (who came to Australia at least 40 000 years ago and could not have introduced the animal). It is an inhabitant of woodland and grassland, often the edge of forest, feeding on the Rabbit, a wide range of terrestrial marsupials, rodents, reptiles and sheep. It hunts in packs for large prey, singly when feeding on small animals. Mating occurs from autumn to early winter and litters of three or four are born from late winter to spring. Dingoes interbreed freely with feral domestic Dogs and few undiluted populations remain. There is compelling circumstantial evidence that the Dingo was responsible for the extermination of the Thylacine and Tasmanian Devil on the Australian mainland.

Australian Fur-seal *Arctocephalus pusillus* TL 1.25-2.25m

R.M. Warneke

This genus comprises eight species, two off Australian coasts. They have small ears and differ from sea-lions in having a much denser underfur. The male Australian Fur-seal is dark brown all over except for a slightly paler mane. The female, which is much smaller, is silvery-grey above, with a creamy throat. It feeds, often at great depth, on squids, fishes and bottom-living invertebrates. For most of the year, a rookery is occupied only by females and young. The male comes ashore late in October and establishes territory. A single young is born in November or December. Mating occurs about a week later and the male departs about mid-January. The young is suckled for about 12 months.

New Zealand Fur-seal *Arctocephalus forsteri* TL 1.3-2m

N. Brothers

This is smaller than the Australian Fur-seal, dark grey-brown above, merging to a lighter grey-brown below. The male has a thick mane, flecked by pale hairs. The female, which is smaller, is metallic grey to brown above, paler below. It feeds on squids, octopuses, fishes and bottom-dwelling crustaceans. Between March and September, females and sub-adults occupy a rookery. From October to December, the mature male establishes its territory. The pregnant female comes ashore about mid-December and gives birth to a single young after a few days. Mating occurs about 10 days later and the male departs around January. Young are weaned at 10 to 11 months.

Australian Sea-lion *Neophoca cinerea* TL 1.6-2.4m

K.A. Ward

 This genus has one species. The male is blackish to chocolate brown with a prominent white mane. The female is silvery-grey above, creamy below. The snout is blunter than in fur-seals. The diet includes fishes and squids. The pregnant female comes ashore about three days before giving birth to a single young and there is strong evidence of an 18-month breeding cycle. The male establishes its territory and harem and defends these. Mating occurs four to nine days after the young is born. It is suckled for at least 12 months.

Dugong *Dugong dugon* TL up to 3.15m

B. Cropp

With an enormous, bristly upper lip, paddle-like forelimbs, no hindlimbs, and a horizontal tail-fluke, the dugong is unmistakable. Seldom seen except by divers, it inhabits warm, shallow coastal waters and estuaries where sea-grasses grow. Where food is plentiful, it tends to be sedentary, moving into shallower water on the rising tide and retreating as the tide falls, surfacing to breathe every minute or so. It usually swims slowly but can attain a speed of 22 km/h. Where food is sparse it may move up to 25km in a day and range over 100km in a year. It is social, moving in groups of up to 100. Growth is slow and sexual maturity is not reached until the age of 10 to 15 years. A single young is born between September and April at intervals of three to six years and accompanies its mother for one or two years. With such a rate of reproduction, populations cannot withstand heavy human predation, even though some individuals may live to about 50 years.

Glossary

Arboreal. Living in trees.

Arthropods. Animals with jointed legs; insects. spiders, millipedes, etc.

Aggregations. Temporary groupings of animals, usually around a source of food or water.

Breakaway. A flat-topped hill above a plain (in America, a mesa).

Browser. An animal that feeds mainly on leaves of shrub and small trees.

Canopy. The upper foliage of a forest.

Carnivorous. Flesh-eating: often combined with insect-eating (*see* Insectivorous).

Circumpolar. Distributed around the extreme southern or northern seas, including the Atlantic and Pacific Oceans.

Crepuscular. Active around dawn or dusk.

Dentition. The arrangement and shape or the teeth of an animal.

Dimorphic. Occurring in two forms, usually of different colour or size.

Disjunct Distribution. In respect of species, occurring in two or more separate areas.

Distal. Towards the unattached end of an extremity, such as a limb or tail (*see* Proximal).

Dorsal. Pertaining to the back of an animal (*see* Ventral).

Echolocation. Location of objects in the surroundings by reception of echoes of sounds or ultrasounds produced by an animal, as in microbats and seals.

Endemic. Native to a region under discussion.

Epiphyte. A plant (often a fern) that grows on a tree.

Evolutionary radiation. The evolution of an ancestral species (over many millions of years) of many different species, each adapted to different environments or ways of life.

Frosted. Having white or pale tips to fur.

Gestation. The period between conception and birth.

Grazer. An animal that feeds on grasses or similar plants.

Incisor. Teeth at the front of the jaws of a mammal.

Insectivorous. Feeding mainly upon insects or other arthropods (*see* Carnivorous).

Interfemoral membrane. The membrane of skin between the hindlegs of some bats.

Interdigital. Between the fingers or toes.

Mammalogist. A scientist who studies mammals.

Mid-successional. The situation in which regeneration of vegetation after fire is about half-way to completion.

Montane. On, or related to, mountains.

Pitfalls. Traps for small mammals, consisting of a smooth-walled bucket with its lips at ground-level – often with converging barriers that guide terrestrial animals towards its opening.

Pop-hole. One of several (emergency) exits or entrances to a complex burrow.

Prehensile. Capable of grasping, as in a paw or flexible tail.

Proximal. Towards the attached end of an extremity, such as a limb or tail (*see* Distal).

Radiation. *See* Evolutionary Radiation.
Relict. Left over: used for example, in respect of a widespread population now reduced to a limited area.
Rhinarium. A sensitive, often moist, area at the tip of the snout of many mammals.
Understorey. The vegetation on the floor of a forest.
Ventral. Pertaining to the underside of an animal (*see* Dorsal).
Vestigial. Descriptive of an organ or other structure that is so reduced from its presumed ancestral condition – as in the muscles of the human ear.
Volplaning. Gliding.

Suggested further reading

Flannery, T.F. (1994). *Possums of the World.* Geo Productions, Sydney.
King, J.E. (1983). *Seals of the World.* University of Queensland Press, Brisbane.
Larvey, H. (1985). *The Kangaroo Keepers.* University of Queensland Press, Brisbane.
Strahan, R. (1995). *The Australian Museum Complete Book of Australian Mammals.* Reed, Sydney.
Watts, C.H.S. and Aslin, H.J. (1981). *The Rodents of Australia.* Angus & Robertson, Sydney.

Index